D1542996

Progress in Computer Science and Applied Logic
Volume 11

Wayne Snyder

A Proof Theory
for General
Unification

1991

Birkhäuser
Boston · Basel · Berlin

Wayne Snyder
Department of Computer Science
Boston University
Boston, MA 02215

Library of Congress Cataloging-in-Publication Data

Snyder, Wayne, 1955-
   A proof theory for general unification / by Wayne Snyder.
        p.   cm. -- (Progress in computer science and applied logic :
   v. 11)
      Includes bibliographical references.
      ISBN 0-8176-3593-9 (hard : acid-free) -- ISBN 3-7643-3593-3
   (hard : acid-free)
      1. Automatic theorem proving.   I. Title.   II. Series.
   QA76.9.A96S6   91-28685
   511.3--dc20   CIP

Printed on acid-free paper.

ISBN 0-8176-3593-9
ISBN 3-7643-3593-9

Camera ready text prepared by the author in LaTeX.
Printed and bound by Edwards Brothers, Inc., Ann Arbor, Michigan.
Printed in the U.S.A.

9 8 7 6 5 4 3 2 1

Table of Contents

Preface

In this monograph we study two generalizations of standard unification, E-unification and higher-order unification, using an abstract approach originated by Herbrand and developed in the case of standard first-order unification by Martelli and Montanari. The formalism presents the unification computation as a set of non-deterministic *transformation rules* for converting a set of equations to be unified into an explicit representation of a unifier (if such exists). This provides an abstract and mathematically elegant means of analysing the properties of unification in various settings by providing a clean separation of the logical issues from the specification of procedural information, and amounts to a set of 'inference rules' for unification, hence the title of this book.

We derive the set of transformations for general E-unification and higher-order unification from an analysis of the sense in which terms are 'the same' after application of a unifying substitution. In both cases, this results in a simple extension of the set of basic transformations given by Herbrand-Martelli-Montanari for standard unification, and shows clearly the basic relationships of the fundamental operations necessary in each case, and thus the underlying structure of the most important classes of term unification problems.

In addition to the presentation of a proof theory for unification which unifies and clarifies the diverse approaches currently being developed for E-unification and higher-order unification, in particular we present the first rigorous analysis of a method for E-unification which is fully general in the sense that it is capable of enumerating a complete set of E-unifiers for arbitrary sets of equations E.

I would like to dedicate this monograph to my wife Jane, for her constant love and encouragement for the last ten years; to Jean Gallier, for his boundless enthusiasm, for teaching me the joy of creative mathematics, and for introducing me to *eau d'vie*; and finally to my parents, Norma and Albert, for all the years of support and love. Thanks to Frank Pfenning, Dale Miller, Paliath Narendran, Dan Dougherty, Patty Johann, and Pierre Lescanne for many useful comments on this work. I would also like to thank Pierre Lescanne and the members of CRIN for another wonderful visit during summer 1991 where I finished the final corrections to this text. Finally, thanks to Phillip Lockhart and our common friend Vergil: *amicus certus in re incerta cernitur*.

CHAPTER 1

INTRODUCTION

Most researchers consider the modern period of automated logic to have begun with the discovery of resolution by J.A. Robinson in 1963 at the Argonne National Laboratory. Previously, it was known by the Herbrand-Skolem-Godel theorem that semi-decision procedures could be designed for first-order logic by reducing the question of the unsatisfiability of a set of first-order formulae to the question of unsatisfiability of (roughly) a set of certain ground formulae derived from the original set in an effective way (for example, see [32]). But until Robinson invented the simple and powerful inference rule known as *resolution* [139], no practically efficient semi-decision procedure had been found. The crucial component of this seminal discovery was in fact the rediscovery by Robinson of the process of *unification*, which had been discovered by Herbrand in his thesis 33 years earlier (see Appendix 3).[1]

The unification problem is the following: *Given two first-order terms s and t, does there exist a substitution θ of terms for the variables in s and t such that $\theta(s) = \theta(t)$?* Robinson [139] showed that this problem is decidable and that whenever a solution, or *unifier*, exists, there always exists a *most general unifier* (or *mgu*) from which all other solutions can be generated, and which is unique in a certain sense (see Section §3.3). After Herbrand and Robinson, this problem was studied extensively by various researchers [27,30,72,109,122,142,155] and, among other results, it was shown that linear time algorithms for unification exist [110,122].

Undoubtedly unification has become familiar to most computer scientists through the revolutionary development of a restricted form of resolution (SLD-resolution) into a programming language, *Prolog*, in the early 1970's by A. Colmerauer, R. Kowalski and others [29,106,95]. But it has also found applications in generalizations of functional programming languages [105, 134], in type inferencing for polymorphic programming languages

---

[1] As an interesting sidelight, we should remark that Dag Prawitz in 1960 revived the notion of unification and used it in a theorem proving method on a computer [132], and in 1964 J.R. Guard quite independently reported on a project in semi-automated mathematics which used unification in its inference rules [60].

[116] (but see also [89]), in expert systems [147], and in the calculation of critical pairs in the Knuth-Bendix completion procedure [94], which is the most important theorem proving method yet developed in equational logic. In addition, Peter Andrews invented a new theorem proving method very different from resolution, called *matings*, which also uses unification as a central operation [4] (a dual version of this was developed independently by W. Bibel [19]). Thus it is apparent that unification is fundamental to many diverse aspects of computational logic.

Since the early 1960's there have been many attempts to generalize the basic paradigm of theorem proving (and, more recently, logic programming), and these have stimulated research into more general forms of unification. The two most significant developments in this regard are the incorporation of equality into theorem proving procedures, and the attempts to automate higher-order logic. Each of these has spawned a new kind of unification.

The prevalence of equational reasoning, i.e., the "substitution of equals for equals" in ordinary mathematical reasoning, and the expressive power of first-order logic to define algebraic structures convinced many researchers to attempt to introduce equality into resolution [2,14,15,21,38,68,100,101,117, 124,137,138,143,146,152,160]. In view of the inherent inefficiency of these methods, Robinson and then Plotkin [131] suggested that when the set of clauses can be partitioned into a set of equational axioms and a set of standard (non-equational) clauses, theorem provers should be stratified into a (non-equational) refutation mechanism and an *E-unification* mechanism, which performs equational reasoning only during unification steps.

Given a (finite) set E of equations and two terms u and v, a substitution θ is called an *E-unifier of u and v* iff $\theta(u)$ and $\theta(v)$ are provably equal under the equations in E, that is, congruent modulo the least stable congruence $\xleftrightarrow{*}_E$ containing E (see Section §4.1). Unlike standard unification (i.e., where $E = \emptyset$), E-unification is undecidable in general, due to the undecidability of the word problem for semigroups. Another major difference is that if u and v are E-unifiable there may not be a single *mgu*, but (possibly) an infinite set of unrelated E-unifiers. Thus the completeness of E-unification procedures must be defined in terms of *complete sets of E-unifiers*.

E-unification is important not only in pure theorem proving, where it allows one to automate reasoning about various algebraic structures defined by sets of equations, but also in the context of logic programming, where it provides a theoretical basis for the incorporation of functional and equational languages into the basic paradigm as represented, for example, by Prolog [50, 57, 133]. This research has received much attention in the last fifteen years, primarily due to the development of the Knuth-Bendix completion procedure. To date, many special purpose E-unification procedures have been defined for particular equational theories (for a good summary see [145]), and several general algorithms [22,46,48,80,78,91,111,120, 136,135,158] for the class of theories which can be compiled into a canonical set of rewrite rules. Unfortunately, the general problem of E-unification for arbitrary theories has been little studied, and there has been no attempt, as far as we know, to define a computationally feasible procedure for general E-unification separate from a refutation procedure. In general there seems to be a need for some integrated approach which will show the structure of the class of all E-unification problems.

Higher-order unification is a method for unifying terms in the Simple Theory of Types [28], that is, given two typed lambda-terms e_1 and e_2, finding a substitution σ of lambda-terms for the free variables of e_1 and e_2 such that $\sigma(e_1)$ and $\sigma(e_2)$ are equivalent under the conversion rules of the calculus. Like E-unification, higher-order unification is undecidable and *mgus* may not exist, requiring the consideration of complete sets of higher-order unifiers.

This problem is fundamental to automating higher-order reasoning, as convincingly shown for example in the automated proof of Cantor's Theorem (that there is no surjection from a set to its powerset) found by the TPS system [5], where the higher-order unification procedure finds a term which corresponds to the diagonal set $\{a \in A \mid a \notin f(a)\}$ used in the standard proof (for details, see [5]). Higher-order unification has formed the basis for generalizations of the resolution principle to second-order logic [31,128] and general ω-order logic [69,123,129] (but see also [3]), the generalization of the method of matings [4] to higher-order [6,5, 112,125], higher-order logic programming in the language λProlog [113,118], a means for providing flexible implementations of logical inference rules in theorem provers [47,123], program synthesis, transformation, and development [74,62,63, 114,127], and also has applications to type inferencing in polymorphic languages [126], computational linguistics [115], and certain problems in proof theory concerning the lengths of proofs [44]. Higher-order unification was studied by a number of researchers [31,59,60,61, 128,129] before Huet [71,72] made a major contribution in showing that a restricted form of unification, called preunification, is sufficient for most refutation methods and in defining a method for solving this restricted problem which is used by most current higher-order systems.

In this monograph, we study these two generalizations of unification using the formalism described by Herbrand for standard unification, and developed by Martelli and Montanari [109]. This method, which will be presented in Section §3.3 and developed in Chapters §5 through §7, presents the unification computation as a non-deterministic set of transformations on *systems of equations*, and amounts to an inference system for transforming a unification problem into an explicit representation of its own solution, if such exists. This method—which has certain similarities to the method of Gaussian Elimination in linear algebra—provides an abstract and mathematically elegant means of analyzing the invariant properties of unification in various settings by providing a clean separation of the logical issues from the specification of procedural information.[2] In addition, this abstract approach clarifies the fundamental mathematical properties of the procedure, such as termination and completeness. Using this method we were able to prove for the first time the completeness of a general E-unification method.

But we also intend this approach to be a formalism for studying the basic properties of unification in its various guises, to form the basis for a "proof theory of unification." By presenting the unification computation as a set of basic abstract operations on equation systems, we are able to isolate and study the significance of these fundamental operations in various settings, and to understand something of the structure of the class of all unification problems. The method for investigating these generalizations of unification is the same in each case: by a close examination of what it means for two terms to become 'the same' after application of a unifying substitution, we derive the appropriate extension to the basic set of transformations ST for standard unification (given in Section §3.3). We then use this abstract formalism to prove the appropriate soundness and completeness results, and then see what restrictions can be made on the basic set to improve efficiency while retaining completeness.

In the case of E-unification, the analysis of the relation $\xleftrightarrow{\;*\;}_E$ suggests two different sets of transformations BT and T which account for the presence of arbitrary equations in a unification problem. These sets are proved to be complete in the sense that for every set E of equations, a complete set of E-unifiers can be enumerated using transformations from either of these sets. The set T is an improvement of BT, however, in that many redundant E-unifiers produced by BT will be weeded out by T. Although BT only contains two more transformations than ST, and T one more transformation than ST, proving the completeness of BT and T turned out to be quite difficult. We were led to define a new representation of equational proofs as certain kinds of (sets of) trees. These proof trees are used to prove the completeness of the set BT in a rather direct fashion that parallels the completeness of the simple set ST in the case of (standard) unification. In order to prove the completeness of T, inspired by the concept of *unfailing*

---

[2] Cf. Kowalski's famous dictum from [96]: "Algorithm = Logic + Control".

completion [8,12,9], we developed an abstract (and simpler) notion of the completion of a set of equations that allowed us to use the previous completeness proof. We also use this abstract form of completion to prove the completeness of a generalization of narrowing (or surreduction) and then give a second proof of the completeness of T based on the completeness of the generalization of surreduction. Finally, we prove the completeness of a refined version T' of the inference rules based on the technique of Relaxed Narrowing (which it itself a refinement of Lazy Paramodulation).

This part of the book generalizes the approach initiated in the pioneering work of Kirchner [91] to arbitrary theories. One of the main important technical differences between our work and Claude Kirchner's is that we use transformations extending naturally those proposed by Herbrand [64], whereas Kirchner uses transformations closer to those Martelli and Montanari developed for multiequations [109]. Also, Kirchner's transformations are only complete for a subclass of all equational theories, the strict theories. Nevertheless, our work would not have been possible without Claude Kirchner's previous contribution. Another concept that inspired us at a crucial time is the idea of unfailing completion, due to Bachmair, Dershowitz, Hsiang, and Plaisted [8,12,9]. Without this research, we would not have been able to show the completeness of our improved set of transformations T.

In the case of higher-order unification, we present two sets of transformations, $\mathcal{HT}$ and $\mathcal{PT}$, which are developed from an analysis of the manner in which substitution and β-reduction make two terms identical. The set $\mathcal{HT}$ is proved to be complete for arbitrary higher-order terms, but unfortunately, the search space (the tree of all transformation sequences) can be infinitely branching, which forbids a reasonable implementation. The set $\mathcal{PT}$ incorporates Huet's well-known solution to this problem. Our presentation of the higher-order unification problem in this formalism shows clearly the logically invariant properties of first-order unification, higher-order preunification, and general higher-order unification. Our major contribution in this chapter of the monograph is three-fold. First, we have extended the Herbrand-Martelli-Montanari method of transformations on systems to higher-order unification and pre-unification; second, we have used this formalism to provide a more direct proof of the completeness of a method for higher-order unification than has previously been available; and, finally, we have shown the completeness of the strategy of eager variable elimination. In addition, this analysis provides another justification of the design of Huet's procedure, and shows how its basic principles work in a more general setting.

The plan of this book is as follows. In Chapter §2 we provide an overview of the method of transformations for unification in order to motivate their use as a general framework for investigating unification problems. In Chapter §3, we present a comprehensive introduction to the the major definitions and results which form the background to the results in the later chapters, including a detailed presentation in Section §3.3 of the Herbrand-Martelli-Montanari method for standard unification via transformations on systems of equations. The soundness and completeness results in this section form the basic paradigm under which we shall study the more general sets of transformations later in the book. The next three chapters present our results on general E-unification. Chapter §4 outlines the basic results and definitions relating to E-unification, and contains a detailed proof of the completeness of the method of *narrowing*, which is the most general form of E-unification investigated to date. Chapter §5 presents the set of transformations $\mathcal{BT}$ and proves its soundness and completeness for arbitrary E using a new representation for proofs of E-unifiability, known as *equational proof trees*. Chapter §6 contains the improved set of transformations $\mathcal{T}$ and the associated soundness and completeness results. We conclude this chapter with a comparison with previous work done on more general forms of E-unification and a discussion of an open problem regarding the use of one of the transformation rules, known as *variable elimination*. In Chapter §7 we extend the Herbrand-Martelli-Montanari method to higher-order unification. After presenting a detailed review of the basic notions of the typed lambda-calculus, the conversion rules, and higher-order substitutions, we present the two sets of transformations $\mathcal{HT}$ and $\mathcal{PT}$ and prove their soundness and completeness. Chapter §8 summarizes our major results. The appendices contain material felt to be incidental to the main thread of the monograph.

CHAPTER 2

PREVIEW

The method of transformations for solving unification problems is much like the well-known method used for solving systems of linear equations known as Gaussian elimination. In Gaussian elimination, the original system of equations is transformed step by step (by variable elimination) into a solved system, that is, a system whose solution is obvious. Similarly, a unification problem is a set $\{u_1 \approx v_1, \ldots, u_n \approx v_n\}$ of equations between terms (sometimes called a *disagreement set*) to be (jointly) unified. (We consider these equations to be *unoriented*.) The method of transformations consists of applying simple transformations, some akin to variable elimination, until a "solved" system S' is obtained whose solution is obvious (in a sense to be made precise below).

Gaussian elimination and first-order unification are somewhat similar. For example, the transformations for first-order unification given in Section §3.3, like Gaussian elimination, must terminate and hence the existence of solutions is decidable. Also, these transformations preserve the set of solutions as an invariant, just as in Gaussian elimination the variable elimination step preserves solutions; and in both the set of solutions is either empty, a singleton, or infinite. But for higher-order unification and E-unification the analogy breaks down. For example, unlike Gaussian elimination, it is undecidable whether a higher-order system has unifiers or whether a first-order system has E-unifiers, and the transformations do not terminate in general. Also, the transformations used for these more complex forms of unification do not necessarily preserve the set of solutions. In general, if a system S' is derived from a system S, it can only be claimed that the set of unifiers of S' is a subset of the set of unifiers of S. Thus, we face a completeness problem: we have to show that every unifier of S will be produced as the obvious solution of some system S' derivable from S. In fact, it is practically impossible to require that every unifier of S be produced, and normally we are only interested in whether a complete set of unifiers can be enumerated using the transformations. Roughly speaking, a complete set of unifiers for S is a set of unifiers for S from which every unifier for S can be generated.

Thus the interesting issue is in finding natural sets of transformations which present in an abstract form the fundamental operations of unification, but which are complete in this sense. In order to introduce the notion of E-unification and of higher-order unification, we shall first demonstrate the full method in the first-order case, and then sketch what changes need to be made to deal with equations on the one hand and with higher-order terms on the other. This will hopefully provide the necessary intuition for the more detailed treatment in the remainder of the monograph.

Suppose we wish to find a unifier for the two terms $f(x, f(h(x, gx), x'))$ and $f(x, f(h(fy, z), y'))$. Now any substitution which unifies these terms can not affect the topmost function symbol f, and so it is easy to see that a substitution θ unifies the terms if and only if it pairwise unifies each of the immediate subterms. For example, θ unifies the system

$$\{f(x, f(h(x, gx), x')) \approx f(x, f(h(fy, z), y'))\}$$

iff it unifies

$$\{x \approx x, f(h(x, gx), x') \approx f(h(fy, z), y')\}.$$

In general, we may define a transformation on systems which we call *term decomposition*:

$$\{f(u_1, \ldots, u_n) \approx f(v_1, \ldots, v_n)\} \cup S \implies \{u_1 \approx v_1, \ldots, u_n \approx v_n\} \cup S,$$

where S is any system (possibly empty). After two more iterations of this transformation, we have

$$\{x \approx x, x \approx fy, gx \approx z, x' \approx y'\}.$$

Now in this system, it is clear that the equation $x \approx x$ is in fact already unified, and contributes no information about possible solutions, since *any* substitution unifies an equation $u \approx u$ for some term u. Thus we may define a transformation which simply removes such trivial equations:

$$\{u \approx u\} \cup S \implies S.$$

In our example, we may derive the new system

$$\{x \approx fy, gx \approx z, x' \approx y'\}.$$

These two transformations simplify a system (by reducing the total number of symbols in the whole system) but do not in any way change the set of solutions; hence the set of solutions is *invariant* under the transformations. But it is not yet obvious what the set of solutions is. The reader may check for example that $[fy/x, gfy/z, x'/y']$, $[fy/x, gfy/z, y'/x']$, and $[fha/x, gfha/z, ha/y, a/x', a/y']$ are all unifiers of the system. In each of these however, the binding made for x has the form ft for some term t, since if a substitution unifies the equation $x \approx fy$ then the binding for x must have f as a top symbol. In this case, we may provide a partial binding for x (since we do not yet know the entire binding, but only the top symbol) by transforming the previous system into a new one which contains this partial binding:

$$\{x \approx fx_1, x \approx fy, gx \approx z, x' \approx y'\}.$$

Now we may eliminate the variable x from the rest of the system by replacing it by fx_1, i.e., by applying the substitution $[fx_1/x]$. After applying decomposition once more, we get the system

$$\{x \approx fx_1, x_1 \approx y, gfx_1 \approx z, x' \approx y'\}.$$

In general, we may define an *imitation rule* for partially solving variables in systems: If x does not occur in the term $f(t_1, \ldots, t_n)$ then we have:

$$\{x \approx f(t_1, \ldots, t_n)\} \cup S \implies$$
$$\{x \approx f(y_1, \ldots, y_n), f(y_1, \ldots, y_n) \approx f(t_1, \ldots, t_n)\} \cup S',$$

where $y_1, \ldots, y_n$ are *new* variables occurring nowhere else, and S' is the result of replacing every occurrence of x in S by the partial binding $f(y_1, \ldots, y_n)$. Note that if x were to occur in the term $f(t_1, \ldots, t_n)$ then the system would not be unifiable.

The point of the imitation rule is that we find a partial solution for a variable x, and then *solve* x partially by substituting the partial solution for the remaining occurrences of x, thus reduced the problem of finding a binding to solving for the new variables in the partial binding for x. Intuitively, if we transform a system using the rule

$$\{x \approx t\} \cup S \implies \{x \approx t\} \cup S[t/x],$$

where x is a variable occurring in S but not occurring in t and $S[t/x]$ represents the result of replacing every occurrence of x in S by t, then, as in Gaussian Elimination, we have *solved* the system for the variable x; hence this transformation is called *variable elimination*. As in the case of our first two transformations, the set of solutions is invariant under variable elimination. (Imitation does not preserve solutions, since it potentially introduces new variables.) In our example, we can eliminate the variable x_1 to obtain the system

$$\{x \approx fy, x_1 \approx y, gfy \approx z, x' \approx y'\}.$$

If we say that an equation $x \approx t$ is in *solved form* in a system if x does not occur in the rest of the system and does not occur in t, then clearly the last system is *solved* in the sense that all its equations are in solved form.

The basic idea of the transformation method as represented by these four transformations is to successively build up bindings for variables and simplify the systems produced by decomposing and eliminating trivial equations. The intent is to transform a unification problem into a solved system, since a solved system $\{x_1 \approx t_1, \ldots, x_n \approx t_n\}$ gives explicitly the bindings of a unifying substitution $[t_1/x_1, \ldots, t_n/x_n]$. In our example, we have the unifying substitution $[fy/x, y/x_1, gfy/z, x'/y']$, which, since we are only interested in bindings made for the variables in the original system, may be restricted to the form $[fy/x, gfy/z, x'/y']$. (It is interesting to note that we could also have extracted the substitution $[fy/x, gfy/z, y'/x']$.) This set of four transformations can be easily shown to be *sound* in the sense that if $S \implies S'$ and θ unifies S', then θ unifies S; thus the method is correct since any solution found will unify the original system. Showing that the method is *complete* is harder, since we must show that for *any* unifier θ of the original system S, we can find a sequence of transformations $S \overset{*}{\implies} S'$ resulting in a solved form S' such that the substitution $\sigma_{S'}$ extracted from S' is more general than θ (over the set of variables in S). The intuitive reason that we can find *mgu*'s (and, more generally, we can find complete sets in the case of E-unification and higher-order unification) using this method is that imitation and variable elimination are capable of incrementally building up the bindings in the unifying substitution just as much as is necessary to unify the original system. The reader may check for example that each of the substitutions found above for S is more general than *any* unifier of the original system, i.e., they are *most general unifiers* or *mgu*'s.

There are several important things to note about this method. The first is that it is a *non-deterministic* set of abstract operations for unification; we can think of it as a set of *inference rules* for unification. This removal of control and data structure specification allows us to examine the fundamental properties of the problem more clearly. The notion of completeness is also non-deterministic, since we show only that for an arbitrary unifier θ there is *some* sequence of transformations which produces a unifier more general than θ. In order to design a practical procedure, we would have to specify data structures and a search strategy to explore the search tree of possible transformation paths. The second point is that if we need to find *all* unifiers, then in the case of an equation between two variables we would need to apply imitation by 'guessing' a partial binding for one of the variables or by guessing an arbitrary variable as a binding. For example, to find the unifier $[fz/x, fz/y]$ of the system $\{x \approx y\}$ we would have to guess the function symbol f in the imitation equation $x \approx fy_1$, then imitate for y, and finally guess that y_1 is bound to z. This is clearly a problem for implementation, but it turns out that for unification in theorem proving we need only find most general solutions, and so in the first-order case we can avoid this guessing by using variable elimination on such equations. In fact, if we are interested in stopping as soon as the possibility of unification is detected, without necessarily transforming the system into a fully solved form, we may define the notion of a *presolved* system as one consisting of either solved equations, as above, or equations consisting of two variables, and stop the transformation process as soon as a presolved form is reached. For example, the system

$$\{x \approx a, y \approx fz, x' \approx y', x' \approx z', z' \approx x''\}$$

is presolved. It turns out that it is always possible to unify such systems, by applying variable elimination to the variable-variable equations which are not yet solved. This shows that we need never apply the imitation rule to a variable-variable equation, since such equations can always be eliminated using variable elimination; in the higher-order generalization of this case, this is not true, as we shall see, and the notion of presolved forms is crucial. It is interesting that in first-order, the presence of variable-variable equations is the reason that *mgu*'s are, strictly speaking, not unique; recall that in our example above, we had two choices about the extraction of a binding from the equation $x' \approx y'$, resulting in the two *mgu*'s $[fy/x, gfy/z, x'/y']$ and $[fy/x, gfy/z, y'/x']$.

The other interesting point is that in the first-order case we have presented, we can in fact have a complete set of transformation rules if we exclude the imitation rule, i.e., if we find bindings by simply eliminating a variable all at once if we find an equation $x \approx t$ where x does not occur in t. In our previous example, we could have 'short-circuited' the sequence of transformations by immediately eliminating the variable x to produce a solved form:

$$\{f(x, f(h(x, gx), x')) \approx f(x, f(h(fy, z), y'))\}$$
$$\overset{*}{\Longrightarrow} \{x \approx fy, gx \approx z, x' \approx y'\}$$
$$\Longrightarrow \{x \approx fy, gfy \approx z, x' \approx y'\}.$$

In Section §3.3 we shall develop this improved method in detail; the completeness of these transformations is particularly easy to prove.

The method we have just sketched can be generalized to E-unification by adding just a single new transformation to account for the presence of an equational theory (that is, we attempt to solve a system, i.e., a set of equations, wrt another set of equations representing an equational theory). The basic idea is that if some substitution θ unifies two terms s and t modulo a set of equations E, then there must exist a sequence of rewrite steps (to be defined precisely in the next chapter) between the two terms $\theta(s)$ and $\theta(t)$:

$$\theta(s) = s_0 \longleftrightarrow_E s_1 \longleftrightarrow_E s_2 \longleftrightarrow_E \ldots \longleftrightarrow_E s_n = \theta(t).$$

For example the substitution $\theta = [a/x, b/y]$ unifies the terms $s = f(x)$ and $t = g(y)$ modulo the set of equations $E = \{f(a) \doteq g(b)\}$, since

$$\theta(s) = f(a) \longleftrightarrow_{[f(a) \doteq g(b)]} g(b) = \theta(t),$$

where the notation $\longleftrightarrow_{[f(a) \doteq g(b)]}$ simply indicates that we have rewritten $f(a)$ to $g(b)$ (we are simplifying here, since in general, equations will have variables). We must define a new transformation which allows us to simulate these rewrite steps. Since the equational proof step replaced a term at the root of $\theta(s)$ by another term, we can think of reducing the problem of E-unifying the system $\{f(x) \approx g(y)\}$ to the problem of E-unifying the new system $\{f(x) \approx f(a), g(b) \approx g(y)\}$, that is, we must try to make s look like one side of the equation, and t like the other. After that, we can apply decomposition to each equation to obtain the solution $\{x \approx a, b \approx y\}$. This motivates the new transformation which we call *root rewriting*:

$$\{s \approx t\} \cup S \implies \{s \approx l, r \approx t\} \cup S,$$

where S is any system (possibly empty) and either $l \doteq r$ or $r \doteq l$ is an equation from E. The point is that we try to simulate rewrite steps between the two terms which occur at the root. The reader may then wonder what happens when rewrite steps occur below the root of s and t. If both terms are compound, e.g., $s = f(s_1, \ldots, s_n)$ and $t = f(t_1, \ldots, t_n)$, then we can use term decomposition to break this problem into the n subproblems of E-unifying s_1 with t_1, s_2 with t_2, etc. Since no rewrites occur at the root, all the rewrite steps occur between these subterms and can then be discovered separately. But what if no rewrite occurs at the root and one of the terms is a variable? For example, $\theta = [f(a)/x]$ is a unifier of the terms $s = f(g(x))$ and $t = x$ modulo the set of equations $E = \{g(f(a)) \doteq a\}$, that is,

$$\theta(s) = f(g(f(a))) \longleftrightarrow_{[g(f(a)) \doteq a]} f(a) = \theta(x),$$

but the replacement of $g(f(a))$ by a does *not* take place at the root. It happens that in this case, we can use the imitation rule, since if no rewrite takes place at the root, then the binding for x must be in the form $f(u)$ for some u. Thus, we transform the system $\{f(g(x)) \approx x\}$ into

$$\{x \approx f(y), f(g(f(y))) \approx f(y)\}$$

and then, by decomposition, into

$$\{x \approx f(y), g(f(y)) \approx y\}.$$

Now we can apply root rewriting to obtain

$$\{x \approx f(y), g(f(y)) \approx g(f(a)), a \approx y\},$$

which, after decomposition, becomes

$$\{x \approx f(y), y \approx a\}$$

(where we suppress the redundant equation $a \approx y$, since for the purposes of this preview we defined a system as a *set*). An application of variable elimination produces the solved system $\{x \approx f(a), y \approx a\}$, from which the solution $[f(a)/x]$ can be extracted.

It turns out that by adding this rule to the four already given, we have a complete set of transformations for E-unification for any arbitrary set of equations E. There are several additional complications in the equational case, however. The first is that it is undecidable if two terms are E-unifiable, and so the transformations may not terminate. The second is that most general E-unifiers may not exist, and so we must define the completeness of the method in terms of most general *sets* of unifiers, called *complete sets of unifiers* (which may be infinite!). A final complication is that in using this set of transformations it is not obvious that we can avoid the 'guessing' of partial bindings in the case of equations consisting of two variables, as discussed in the non-equational case above, and so a naive implementation of this method would be impractical. Fortunately, we can show that in the case of E-unification, this guessing of bindings can be avoided without sacrificing completeness, and in a later part of the monograph we shall present a set of transformations in which variable elimination is always applied to such equations, and we shall prove its completeness.

The extension of the original Herbrand transformations to higher-order unification can similarly be made with relatively few changes. The most important differences have to do with the imitation rule and the generalization of the notion of a partial binding to higher-order substitutions. Consider the system $S = \{F(f(a)) \approx f(F(a))\}$, where F is a variable of functional type (say $int \to int$). It is easily seen that $\theta = [\lambda x.f(x)/F]$ is a unifier for S, since

$$\theta(F(f(a))) = (\lambda x.f(x))\,f(a)$$
$$\longrightarrow_\beta \; f(f(a))$$
$$\longleftarrow_\beta \; f((\lambda x.f(x))a) = \theta(f(F(a))),$$

where $\longrightarrow_\beta$ denotes β-reduction. (This is not the only solution, for example the reader may check that any substitution in the form $[\lambda x.\,f^k(x)/F]$ for $k \geq 0$ is also a unifier.) This time, it is a little more tricky to build up θ using partial bindings. In the first-order case, we generate bindings of the form $[f(y_1, \ldots, y_n)/x]$, where $y_1, \ldots, y_n$ are first-order variables. The generalization (roughly) is to consider partial bindings of the form

$$[\lambda x_1 \ldots x_k.\, a\big(Y_1(x_1, \ldots, x_k), \ldots, Y_n(x_1, \ldots, x_k)\big)/F],$$

where $Y_1, \ldots, Y_n$ are some higher-order variables of appropriate types and a is an atom (i.e., a constant, a free variable, or a bound variable x_i for $1 \leq i \leq k$). The idea is that we have to generalize the partial binding $f(y_1, \ldots, y_n)$ to higher-order, and so the top function symbol a may be a variable, and the variables $y_1, \ldots, y_n$ and the term itself may be of functional type; furthermore, each y_i must be generalized to a term $Y_i(x_1, \ldots, x_n)$ since the subterms of the binding may be some function of the bound variables $x_1, \ldots, x_n$. A further level of complexity is introduced by the constraints imposed by the type structure. The notion of higher-order partial bindings will be carefully defined in Chapter 7.

The imitation rule must accommodate this more complex form of partial binding. In the first order case, we applied imitation to an equation $x \approx f(t_1, \ldots, t_k)$ using a partial binding $f(y_1, \ldots, y_n)$; in the higher-order case we must be able to apply imitation to equations such as $F(f(a)) \approx f(F(a))$ to partially solve for F. A partial binding for F which imitates the symbol f in this case would have the form $\lambda x.\, f(Y(x))$, so that we would transform the system $\{F(f(a)) \approx f(F(a))\}$ into

$$\{F \approx \lambda x.\, f(Y(x)), f(Y(f(a))) \approx f(f(Y(a)))\}$$

using the imitation rule; note that we have performed β-reduction after applying the substitution $[\lambda x.\, f(Y(x))/F]$. After decomposition we have

$$\{F \approx \lambda x.\, f(Y(x)), Y(f(a)) \approx f(Y(a))\}.$$

Unfortunately, the imitation rule alone is not sufficient for building up bindings in higher-order unification. This is easy to see in considering the subproblem of finding a partial binding for Y, which is exactly the problem we faced with F; simply continuing to imitate will produce an infinite sequence of transformations. The problem arises because higher-order terms may have variables as their top-most symbol and so we must allow bindings such as $\lambda x.x$ to be found by our transformations. If we abbreviate a lambda binder $\lambda x_1 \ldots x_k$ into the form $\lambda \overline{x_k}$, the new rule for finding partial bindings has (roughly) the form:

$$\{\lambda \overline{x_k}.F(u_1, \ldots, u_n) \approx \lambda \overline{x_k}.a(v_1, \ldots, v_m)\} \cup S \Longrightarrow$$
$$\{F \approx t\} \cup \sigma(\{\lambda \overline{x_k}.F(u_1, \ldots, u_n) \approx \lambda \overline{x_k}.a(v_1, \ldots, v_m)\} \cup S),$$

where a is a function symbol, constant, or variable (either free or bound), and where t is either an imitation binding, i.e.,

$$t = \lambda \overline{y_n}.\, a(Y_1(\overline{y_n}), \ldots, Y_m(\overline{y_n})),$$

or a projection binding, i.e.,

$$t = \lambda \overline{y_n}.\, y_i(Y_1(\overline{y_n}), \ldots, Y_q(\overline{y_n}))$$

for some i, $1 \leq i \leq n$, and $\sigma = [t/F]$ (after applying σ, we also reduce the resulting terms to their normal form using β-conversion). For example, we can transform the system $\{F(f(a)) \approx f(F(a))\}$ by adding a projection binding to get

$$\{F \approx \lambda x.x, F(f(a)) \approx f(F(a))\}$$

and then applying the substitution $[\lambda x.x/F]$ and β-reducing to get

$$\{F \approx \lambda x.x, f(a) \approx f(a)\}.$$

After removing the trivial equation, gives us the solved system $\{F \approx \lambda x.x\}$. The reader may check that a similar projection for the variable Y in our example above results in the solved system $\{F \approx \lambda x.\, f(x), Y \approx \lambda x.x\}$.

Besides the more complicated form of the rule which finds partial bindings, there are complications in the higher-order case analogous to those discussed above for E-unification. First of all, unification here is defined modulo the conversion rules of the lambda calculus, so that we shall have to carefully justify our method from an analysis of the means by which substitution and subsequent β-reduction makes terms equal. Another complication is that, as with E-unification, higher-order unification is undecidable in general and most general unifiers do not necessarily exist, so that we must define the notion of completeness in terms of complete sets of unifiers; in fact, the completeness proof is not much harder than in first-order.

A final important difference from the first-order case has to do with the higher-order equivalent of a variable-variable equation of terms, namely, a pair of terms with variables at their heads, e.g. $\lambda x. F(a, x) \approx \lambda x. G(x, a)$. (These are called *flexible-flexible* equations.) Unfortunately, it is not possible to avoid the arbitrary 'guessing' of bindings discussed above and preserve completeness, and so the search tree for unifiers may be infinitely branching. This posed an insurmountable problem for implementation until Huet showed that in the context of a refutation method, it is usually only necessary to determine the possibility of unification, and since such flexible-flexible equations are always unifiable, we can stop after finding a *presolved* form. This restricted form of unification is termed *preunification*.

CHAPTER 3

PRELIMINARIES

This section contains an outline of the major definitions and results related to unification and equational logic, and is basically consistent with [77] and [49]. A separate section of preliminaries relating to higher-order unification will be given in Chapter §7.

We start with the fundamental algebraic definitions of terms and substitutions, and then introduce the basic concepts of (first-order) matching and unification, including a presentation of the method of (standard) unification by transformations on systems of equations. We then present the basic definitions and results in equational logic, rewriting systems, and completion procedures.

3.1 Algebraic Background

We first review some fundamental algebraic notions.

Definition 3.1.1 Let $\mathbf{N}$ be the set of natural numbers. A *ranked alphabet* is a set Σ with an associated function $arity : \Sigma \to \mathbf{N}$ assigning a *rank* or *arity* n to each symbol f in Σ. We denote the set of symbols of arity n by Σ_n. (For example, the set of constants is just Σ_0.)

We may develop the notion of terms explicitly by defining them as (finite) functions over certain sets of integers, known as tree domains.

Definition 3.1.2 Let $\mathbf{N}_+$ denote the set of positive natural numbers and $\mathbf{N}_+^*$ the set of all strings of positive natural numbers, where we denote the empty string by the symbol ϵ. A *tree domain* D is a nonempty subset of $\mathbf{N}_+^*$ satisfying the conditions:

(i) For all $\alpha, \beta \in \mathbf{N}_+^*$, if $\alpha\beta \in D$ then $\alpha \in D$.

(ii) For all $\alpha \in \mathbf{N}_+^*$, for every $i \in \mathbf{N}_+$, if $\alpha i \in D$ then, for every $1 \le j \le i$, $\alpha j \in D$.

For any two tree addresses α and β in some tree domain D, we say that α is an *ancestor* of β, denoted $\alpha \le \beta$, iff α (considered as a string in $\mathbf{N}_+^*$)

is a prefix (not necessarily proper) of β, i.e., there exists some γ such that $\beta = \alpha\gamma$.

Definition 3.1.3 Given a ranked alphabet Σ, a Σ-*tree* is any function $t : D \to \Sigma$ where D is a tree domain denoted by $Dom(t)$ and such that if $\alpha \in Dom(t)$ and $\{i \mid \alpha i \in Dom(t)\} = \{1, \ldots, n\}$, then $t(\alpha) \in \Sigma_n$. We shall denote the symbol $t(\epsilon)$ by $Root(t)$. Given a tree t and some tree address $\alpha \in Dom(t)$, the *subtree of t rooted at* α is the tree, denoted t/α, whose domain is the set $\{\beta \mid \alpha\beta \in Dom(t)\}$ and such that $t/\alpha(\beta) = t(\alpha\beta)$ for all $\beta \in Dom(t/\alpha)$.[1] Given two trees t_1 and t_2 and a tree address $\alpha \in Dom(t_1)$, the *result of replacing* t_2 at α in t_1, denoted $t_1[\alpha \leftarrow t_2]$, is the function whose graph is the set of pairs

$$\{(\beta, t_1(\beta)) \mid \alpha \not\leq \beta \text{ and } \beta \in Dom(t_1)\} \quad \cup \quad \{(\alpha\beta, t_2(\beta)) \mid \beta \in Dom(t_2)\}.$$

We shall denote the *depth* of a tree t, i.e., the length of the longest path in t (or, equivalently, the length of the longest string in $Dom(t)$), by $|t|$. For example, $|f(a)| = 1$ and $|x| = 0$. The *size* of a tree will be the number of symbols (i.e., the cardinality of $Dom(t)$), and will be denoted by $\|t\|$. The set of all finite Σ-trees (i.e. trees with finite domains) is denoted by T_Σ.

It will be useful to use the notation $t[s]$ to indicate that the tree t contains a subtree s, and more generally, to use $t[s_1, \ldots, s_n]$ to indicate the presence of subtrees $s_1, \ldots, s_n$.

Definition 3.1.4 Given a ranked alphabet Σ, a Σ-*algebra* $\mathcal{A}$ is a pair (A, I), where A is a non-empty set, called the *carrier*, and I is an *interpretation function* assigning functions to the function symbols in Σ such that if $f \in \Sigma_n$, then the symbol f is interpreted as some function $I(f) : A^n \to A$; in particular, when $n = 0$ (i.e., f is a constant), we have $I(f) \in A$. For a given algebra $\mathcal{A}$, we shall denote the function $I(f)$ by $f_\mathcal{A}$.

For any alphabet Σ there exists a special kind of algebra, variously called the *free Σ-algebra*, the *term algebra*, or the *initial algebra*.

Definition 3.1.5 The *free Σ-algebra* is the algebra (T_Σ, I) where for every $f \in \Sigma_n$ with $n \geq 0$ and for all $t_1, \ldots, t_n \in T_\Sigma$, $f_{T_\Sigma}(t_1, \ldots, t_n)$ is the tree denoted $f(t_1, \ldots, t_n)$.

---

[1] Note carefully that t/α is the *subtree* of t at α and $t(\alpha)$ is the *symbol* labelling the node at α.

In other words, the resulting algebra, which will also be denoted by T_Σ, is an algebra of finite trees where the functions in Σ are interpreted as tree constructors. It is well-known that T_Σ is a free algebra (see [49]). To prevent free algebras from having empty carriers, we assume that $\Sigma_0 \neq \emptyset$. It follows that the set T_Σ is nonempty.

The following definition will also be useful.

Definition 3.1.6 Given a Σ-algebra $\mathcal{A}$, a *congruence* $\cong$ on A is an equivalence relation such that, for any $f \in \Sigma_n$ with $n > 0$ and for any $a_1, \ldots, a_n, b_1, \ldots, b_n \in A$, if $a_i \cong b_i$ for $1 \leq i \leq n$, then

$$f_{\mathcal{A}}(a_1, \ldots, a_n) \cong f_{\mathcal{A}}(b_1, \ldots, b_n).$$

Given a countably infinite set of variables $X = \{x_0, x_1, \ldots\}$, we can form the Σ-algebra $T_\Sigma(X)$ by adjoining the set X to the set Σ_0. Thus, $T_\Sigma(X)$ is the set of all finite trees formed from the constant and function symbols in Σ and the variables in X. In order that $T_\Sigma(X)$ be non-empty, we assume that $\Sigma_0 \cup X \neq \emptyset$. It is well-known that $T_\Sigma(X)$ is the free Σ-algebra generated by X (see [49]). This property will allow us to define substitutions.

Definition 3.1.7 A *term* is any member of $T_\Sigma(X)$. The set of variables occurring in a term t is the set

$$Var(t) = \{x \in X \mid t(\alpha) = x \text{ for some } \alpha \in Dom(t)\}.$$

Any term t for which $Var(t) = \emptyset$ (i.e., a member of T_Σ) is called a *ground term*. The set of variable addresses in $Dom(t)$ is

$$VarDom(t) = \{\alpha \in Dom(t) \mid t(\alpha) \in X\},$$

and the set of non-variable occurrences in $Dom(t)$ is

$$NonVarDom(t) = Dom(t) - VarDom(t).$$

Convention: In the rest of this thesis, we shall use the letters a, b, c, and d to denote constants; f, g, and h to denote functions; x, y, and z to indicate variables; l, r, s, t, u, v, and w for terms; and α, β, γ, and δ for tree addresses.

Additional notational conventions will be introduced as needed, particularly in the introductory material in the chapter on higher-order unification.

Finally, the notion of a *multiset* will be used in several different contexts in this thesis.

Definition 3.1.8 Given a set A, a *multiset* over A is an unordered collection of elements of A which may have multiple occurrences of identical elements. More formally, a multiset over A is a function $M : A \to \mathbf{N}$ (where $\mathbf{N}$ is the set of natural numbers) such that an element a in A has exactly n occurrences in M iff $M(a) = n$. In particular, a does not belong to M when $M(a) = 0$, and we say that $a \in M$ iff $M(a) > 0$. The *union* of two multisets M_1 and M_2, denoted by $M_1 \cup M_2$, is defined as the multiset M such that for all $a \in A$, $M(a) = M_1(a) + M_2(a)$.

To avoid confusion between multisets and sets, we shall always state carefully when an object is considered to be a multiset. Note carefully that multiset union is a distinct notion from the union of sets.

3.2 Substitutions

The notion of a *substitution* is central to unification and matching problems.

Definition 3.2.1 A *substitution* is any function $\theta : X \to T_\Sigma(X)$ such that $\theta(x) \neq x$ for only finitely many $x \in X$.

Since $T_\Sigma(X)$ is the free Σ-algebra generated by X, every substitution $\theta : X \to T_\Sigma(X)$ has a unique homomorphic extension $\widehat{\theta} : T_\Sigma(X) \to T_\Sigma(X)$ (see [49]). In the sequel, we will identify θ and its homomorphic extension $\widehat{\theta}$. Less formally, we have the recursive definition

$$\widehat{\theta}(t) = \begin{cases} \theta(x), & \text{if } t = x \text{ for some } x \in X\,; \\ f(\widehat{\theta}(t_1), \ldots, \widehat{\theta}(t_n)), & \text{otherwise, for } t = f(t_1, \ldots, t_n)\,, \end{cases}$$

for some function symbol f and terms $t_1, \ldots, t_n$, where $n \geq 0$.

In the rest of this thesis, we use the greek letters θ, σ, η, ρ, and φ to denote substitutions.

Definition 3.2.2 Given a substitution σ, the *support* (or *domain*) of σ is the set of variables

$$D(\sigma) = \{x \mid \sigma(x) \neq x\}.$$

A substitution whose support is empty is termed the *identity substitution*, and is denoted by Id. The set of variables *introduced by* σ is the set of variables

$$I(\sigma) = \bigcup_{x \in D(\sigma)} Var(\sigma(x)).$$

Given a substitution σ, if its support is the set $\{x_1, \ldots, x_n\}$, and if $t_i = \sigma(x_i)$ for $1 \leq i \leq n$, then σ is also denoted by $[t_1/x_1, \ldots, t_n/x_n]$. Given a term u, we also denote $\sigma(u)$ as $u[t_1/x_1, \ldots, t_n/x_n]$.

Given a set of 'protected variables' W, a substitution ρ is a *renaming substitution away from* W if $\rho(x)$ is a variable for every $x \in D(\rho)$, $I(\rho) \cap W = \emptyset$, and for every $x, y \in D(\rho)$, $\rho(x) = \rho(y)$ implies that $x = y$. If W is unimportant, then ρ is simply called a *renaming*.

The *restriction* of a substitution θ to a set of variables V, denoted by $\theta|_V$, is defined as the substitution θ' such that

$$\theta'(x) = \begin{cases} \theta(x), & \text{if } x \in V; \\ x, & \text{otherwise.} \end{cases}$$

Definition 3.2.3 The *union* of two substitutions σ and θ, denoted by $\sigma \cup \theta$, is defined by

$$\sigma \cup \theta(x) = \begin{cases} \sigma(x), & \text{if } x \in D(\sigma); \\ \theta(x), & \text{if } x \in D(\theta); \\ x, & \text{otherwise,} \end{cases}$$

and is only defined if $D(\sigma) \cap D(\theta) = \emptyset$.

The *composition* of σ and θ is the substitution denoted by $\sigma \circ \theta$ such that for every variable x we have $\sigma \circ \theta(x) = \widehat{\theta}(\sigma(x))$.

Definition 3.2.4 Given a set V of variables, we say that two substitutions σ and θ are *equal over* V, denoted $\sigma = \theta[V]$ iff $\forall x \in V$, $\sigma(x) = \theta(x)$. We say that σ is *more general than* θ *over* V, denoted by $\sigma \leq \theta[V]$, iff there exists a substitution η such that $\theta = \sigma \circ \eta[V]$. When V is the set of all variables, we drop the notation $[V]$.

Definition 3.2.5 A substitution σ is *idempotent* if $\sigma \circ \sigma = \sigma$.

It is easy to give a necessary and sufficient condition for idempotency.

Lemma 3.2.6 A substitution σ is idempotent iff $I(\sigma) \cap D(\sigma) = \emptyset$.

Proof. By a simple induction on $|t|$, we can show that for any term t, $\sigma(t) = t$ iff $Var(t) \cap D(\sigma) = \emptyset$. Applying this to the term $\sigma(x)$ for each $x \in D(\sigma)$, we see that $\sigma \circ \sigma = \sigma$ iff $\forall x \in X. \sigma(\sigma(x)) = \sigma(x)$ iff $\forall x \in D(\sigma). \sigma(\sigma(x)) = \sigma(x)$ iff $\forall x \in D(\sigma). Var(\sigma(x)) \cap D(\sigma) = \emptyset$ iff $I(\sigma) \cap D(\sigma) = \emptyset$. $\square$

Idempotent substitutions are easier to manipulate and the assumption of idempotency often simplifies a proof. For example, it sometimes turns out to be useful in a proof to 'partially instantiate' a term by an idempotent substitution, a technique which is justified by our next result.

Lemma 3.2.7 If θ is an idempotent substitution and t is a arbitrary term, then $\theta(t) = \theta(t[\alpha \leftarrow \theta(t/\alpha)])$ for any $\alpha \in Dom(t)$.

Proof. We simply observe that

$$\theta(t) = \theta(t)[\alpha \leftarrow \theta(t/\alpha)]$$
$$= \theta(t[\alpha \leftarrow \theta \circ \theta(t/\alpha)])$$
$$= \theta(t[\alpha \leftarrow \theta(t/\alpha)]).$$

$\square$

We now formally justify that we may often restrict our attention to idempotent substitutions without loss of generality, by showing that any substitution is equivalent (over an arbitrary superset of its support) up to renaming with an idempotent substitution.

Lemma 3.2.8 For any substitution σ and set of variables W such that $D(\sigma) \subseteq W$, there exists an idempotent substitution σ' such that $D(\sigma) = D(\sigma')$, $\sigma \leq \sigma'$, and $\sigma' \leq \sigma[W]$.

Proof. Let $D(\sigma) \cap I(\sigma) = \{x_1, \ldots, x_n\}$, let $\{y_1, \ldots, y_n\}$ be a set of *new* variables disjoint from W and $I(\sigma)$, let $\rho_1 = [y_1/x_1, \ldots, y_n/x_n]$, and let $\rho_2 = [x_1/y_1, \ldots, x_n/y_n]$. Now let $\sigma' = \sigma \circ \rho_1$, so that clearly $\sigma \leq \sigma'$ and $D(\sigma) = D(\sigma')$ as required. Since $\rho_1 \circ \rho_2 = Id[W \cup I(\sigma)]$, then $\sigma = \sigma \circ \rho_1 \circ \rho_2 = \sigma' \circ \rho_2[W]$, and thus $\sigma' \leq \sigma[W]$. But by our previous lemma, σ' must be idempotent, since $D(\sigma') = D(\sigma)$ is disjoint from $I(\sigma') = (I(\sigma) - \{x_1, \ldots, x_n\}) \cup \{y_1, \ldots, y_n\}$. $\square$

Since most uses of substitutions in this thesis are modulo renaming, this lemma will allow us to assume that substitutions are idempotent if necessary. We shall prove specific results related to the use of idempotent unifiers in later sections.

Finally, we define the notion of a substitution instance and a matching substitution.

Definition 3.2.9 For any term t, a term $\theta(t)$ is called a *substitution instance of* t. Given two terms s and t, a substitution σ is called a *matching substitution* of s and t iff $s = \sigma(t)$. We say that t has been matched to s. (Our convention will be say that two terms can be matched if the second can be matched to the first.)

In other words, two terms match if the first is a substitution instance of the second. The associated decision problem (i.e., whether such a σ exists) is called the *matching problem* for s and t.[2]

---

[2] We remark that $s = \sigma(t)$ does not in itself guarantee that $D(\sigma) \subseteq Var(t)$, so that

3.3 Unification by Transformations on Systems

We now define unification of terms and present an abstract view of the unification process as a set of non-deterministic rules for transforming a unification problem into an explicit representation of its solution, if such exists; in Chapters §5 and §6 this will be extended to E-unification, and in Chapter §7 to higher-order unification. This elegant approach is due to [109], but was implicit in Herbrand's thesis [64] (see Appendix Three, where we quote the passage in full). Our representation for unification problems is the following.

Definition 3.3.1 An *equation* is a pair of terms, denoted, e.g., by $s \doteq t$, and if $Var(s,t) = \emptyset$ then $s \doteq t$ is called a *ground equation*. We use $s \approx t$ to stand (ambiguously) for either $s \doteq t$ or $t \doteq s$ (thus we may think of $s \approx t$ as a multiset $\{s,t\}$ of two terms). A substitution θ is called a *standard unifier* (or just a *unifier*) of an equation $s \approx t$ if $\theta(s) = \theta(t)$. A *equation system* (or just *system*) is a multiset of such equations, and a substitution θ is a unifier of a system if it unifies each equation. The set of unifiers of a system S is denoted $U(S)$, and if S consists of only a single equation $s \approx t$, the set of unifiers is denoted by $U(s,t)$.

Definition 3.3.2 A substitution σ is a *most general unifier*, or *mgu*, of a system S iff
(i) $D(\sigma) \subseteq Var(S)$;
(ii) $\sigma \in U(S)$;
(iii) For every $\theta \in U(S)$, $\sigma \leq \theta$.

It is well known that *mgu*'s always exist for unifiable systems, and it can be shown that *mgu*'s are unique up to composition with a renaming substitution, and so we shall follow the common practice of glossing over this distinction by referring to *the mgu* of a system, denoted by $mgu(S)$.

Definition 3.3.3 An equation $x \approx t$ is in *solved form* in a system S and x in this equation is called a *solved variable* if x is a variable which does not occur anywhere else in S; in particular, $x \notin Var(t)$. A system is in solved form if all its equations are in solved form; a variable is *unsolved* if it occurs in S but is not solved.

---

there may exist some θ more general than σ which also matches t to s. We can, however, assert that for any two matching substitutions σ and σ' of t onto s, it must be the case that $\sigma|_{Var(t)} = \sigma'|_{Var(t)}$. Thus we shall in general assume that if ρ matches t to s, then $D(\rho) \subseteq Var(t)$.

Note that a solved form system is always a *set* of solved equations. The importance of solved form systems is shown by

Lemma 3.3.4 Let $S = \{x_1 \approx t_1, \ldots, x_n \approx t_n\}$ be a system in solved form. If $\sigma = [t_1/x_1, \ldots, t_n/x_n]$, then σ is an idempotent *mgu* of S. Furthermore, for any substitution $\theta \in U(S)$, we have $\theta = \sigma \circ \theta$.

Proof. We simply observe that for any θ, $\theta(x_i) = \theta(t_i) = \theta(\sigma(x_i))$ for $1 \le i \le n$, and $\theta(x) = \theta(\sigma(x))$ otherwise. Clearly σ is an *mgu*, and since $D(\sigma) \cap I(\sigma) = \emptyset$ by the definition of solved forms, it is idempotent. $\square$

Strictly speaking the substitution σ here is ambiguous in the case that there is at least one equation consisting of two solved variables; but since *mgu*'s are considered unique up to renaming, and such equations can be arbitrarily renamed, we denote this substitution by σ_S. As a special case, note that $\sigma_\emptyset = Id$.

We may analyse the process of finding *mgu*'s as follows. If $\theta(u) = \theta(v)$, then either (i) $u = v$; or (ii) $u = f(u_1, \ldots, u_n)$ and $v = f(v_1, \ldots, v_n)$ for some $f \in \Sigma$, and $\theta(u_i) = \theta(v_i)$ for $1 \le i \le n$; or (iii) u is a variable not in $Var(v)$ or vice versa. If u is a variable and $u \notin Var(v)$, then $[v/u] \in U(u, v)$ and $[v/u] \le \theta$. By extending this analysis to account for systems of equations, we have a set of transformations for finding *mgu*'s.

Definition 3.3.5 (The set of transformation rules $\mathcal{ST}$) Let S denote any system (possibly empty), $f \in \Sigma$, and u and v be two terms. We have the following transformations.

Trivial:

$$\{u \approx u\} \cup S \implies S \tag{1}$$

Term Decomposition: For any $f \in \Sigma_n$ for some $n > 0$,

$$\{f(u_1, \ldots, u_n) \approx f(v_1, \ldots, v_n)\} \cup S \implies \{u_1 \approx v_1, \ldots, u_n \approx v_n\} \cup S \tag{2}$$

Variable Elimination:

$$\{x \approx v\} \cup S \implies \{x \approx v\} \cup \sigma(S), \tag{3}$$

where $x \approx v$ is not in solved form, $x \notin Var(v)$, and $\sigma = [v/x]$.

Recall that systems are multisets, so the unions here are multiset unions; the intent of the left-hand side of each of these rules is to isolate a single equation to be transformed. We shall say that $Unify(S) = \theta$ iff there exists some sequence of transformations

$$S \implies \ldots \implies S',$$

where S' is in solved form and $\theta = \sigma_{S'}$. (If no transformation applies, but the system is not in solved form, the procedure given here fails.)

Clearly, by choosing $S = \{u \approx v\}$, we can attempt to find a unifier for two terms u, and v, as the following example shows.[3]

Example 3.3.6

$$f(x, g(a, y)) \approx f(x, g(y, x))$$
$$\implies_{dec} \quad x \approx x, \; g(a, y) \approx g(y, x)$$
$$\implies_{triv} \quad g(a, y) \approx g(y, x)$$
$$\implies_{dec} \quad a \approx y, \; y \approx x$$
$$\implies_{vel} \quad a \approx y, \; a \approx x \, .$$

The sense in which these transformations preserve the logically invariant properties of a unification problem is now shown.

Lemma 3.3.7 If $S \implies S'$ using any transformation from $\mathcal{ST}$, then $U(S) = U(S')$.

Proof. The only difficulty is in Variable Elimination. Suppose $\{x \approx v\} \cup S \implies_{vel} \{x \approx v\} \cup \sigma(S)$ with $\sigma = [v/x]$. For any θ, if $\theta(x) = \theta(v)$, then $\theta = \sigma \circ \theta$, since $\sigma \circ \theta$ differs from θ only at x, but $\theta(x) = \theta(v) = \sigma \circ \theta(x)$. Thus,

$$\theta \in U(\{x \approx v\} \cup S)$$
$$\text{iff} \quad \theta(x) = \theta(v) \; \text{and} \; \theta \in U(S)$$
$$\text{iff} \quad \theta(x) = \theta(v) \; \text{and} \; \sigma \circ \theta \in U(S)$$
$$\text{iff} \quad \theta(x) = \theta(v) \; \text{and} \; \theta \in U(\sigma(S))$$
$$\text{iff} \quad \theta \in U(\{x \approx v\} \cup \sigma(S)).$$

$\square$

The central point here is that the most important feature of a unification problem—its set of solutions—is preserved under these transformations, and hence we are justified in our method of attempting to transform such problems into a trivial (solved) form in which the existence of an *mgu* is evident.

---

[3] In examples, we often drop set brackets for systems, e.g., $S = x_1 \approx t_1, \ldots, x_n \approx t_n$.

We may now show the soundness and completeness of these transformations following [109].

Theorem 3.3.8 (Soundness) If $S \overset{*}{\Longrightarrow} S'$ with S' in solved form, then $\sigma_{S'} \in U(S)$.

Proof. Using the previous lemma and a trivial induction on the length of transformation sequences, we see that $U(S) = U(S')$, and so clearly $\sigma_{S'} \in U(S)$. □

Theorem 3.3.9 (Completeness) Suppose that $\theta \in U(S)$. Then any sequence of transformations

$$S = S_0 \Longrightarrow S_1 \Longrightarrow S_2 \Longrightarrow \ \ldots$$

must eventually terminate in a solved form S' such that $\sigma_{S'} \leq \theta$.

Proof. We first show that every transformation sequence terminates. For any system S, let us define a complexity measure $\mu(S) = \ <n, m>$, where n is the number of unsolved variables and m is the sum of the sizes of all the terms in the system. Then the lexicographic ordering on $<n, m>$ is well-founded,[4] and each transformation produces a new system with a measure strictly smaller under this ordering: Trivial and Term Decomposition must decrease m and can not increase n, and Variable Elimination must decrease n.

Therefore the relation $\Longrightarrow$ is well-founded, and every transformation sequence must end in some system to which no transformation applies. Suppose a given sequence ends in a system S'. Now $\theta \in U(S)$ implies by Lemma 3.3.7 that $\theta \in U(S')$, and so S' can contain no equations of the form $f(t_1, \ldots, t_n) \approx g(t'_1, \ldots, t'_m)$ or of the form $x \approx t$ with $x \in Var(t)$. But since no transformation applies, all equations in S' must be in solved form. Finally, since $\theta \in U(S')$, by Lemma 3.3.4 we must have $\sigma_{S'} \leq \theta$.

□

In fact, we have proved something stronger than necessary in Theorem 3.3.9: it has been shown that all transformation sequences terminate and that *any* sequence of transformations issuing from a unifiable system must eventually result in a solved form. This is possible because the problem is decidable. Strictly speaking, it would have been sufficient for completeness to show that if S is unifiable then there exists *some* sequence of transformations which results in a solved form, since then a complete search strategy,

---

[4] For a definition of the lexicographic ordering, see Definition 3.5.4, and for the notion of well-foundedness, see Definition 3.5.3.

such as breadth-first search, could find the solved form. This form of completeness, which might be termed *non-deterministic completeness*, will be used in finding results on E-unification and higher-order unification, where the general problem is undecidable.

In some contexts it may be useful to deal with idempotent unifiers which are renamed away from some set of 'protected' variables but which are most general over the set of variables in the original system. The next definition makes this precise.

Definition 3.3.10 Given a system S and a finite set V of 'protected' variables, a substitution σ is a *most general unifier of S away from V* (abbreviated $mgu(S)[V]$) iff

   (i) $D(\sigma) \subseteq Var(S)$ and $I(\sigma) \cap (V \cup D(\sigma)) = \emptyset$;
  (ii) $\sigma \in U(S)$;
 (iii) For every $\theta \in U(S)$, $\sigma \leq \theta[Var(S)]$.

That such substitutions may always be found for unifiable systems is shown by

Lemma 3.3.11 If S is a unifiable system and V a protected set of variables, then there exists a substitution σ which is a $mgu(S)[V]$.

Proof. Let $\theta = Unify(S)$, as in Definition 3.3.5, so that θ is an idempotent *mgu* of S such that $D(\theta) \cup I(\theta) \subseteq Var(S)$. If $V \cap I(\theta) = \emptyset$, then $\sigma = \theta$ is a $mgu(S)[V]$. Otherwise, let ρ be a renaming substitution away from $V \cup Var(S)$ such that $D(\rho) = I(\theta)$, and let $\sigma = \theta \circ \rho$. Clearly $D(\sigma) = D(\theta) \cup I(\theta) \subseteq Var(S)$. Since $I(\sigma) = I(\rho)$, by the definition of ρ, σ is idempotent and also $I(\sigma) \cap V = \emptyset$, and hence condition (i) is satisfied. Condition (ii) is satisfied also, since for any equation $u \approx v$ in S, we have that $\theta(u) = \theta(v)$, and thus $\sigma(u) = \rho(\theta(u)) = \rho(\theta(v)) = \sigma(v)$, so that $\sigma \in U(S)$. To show the last condition, we first observe that from the definition of a renaming there must exist an inverse ρ^{-1} such that $\rho \circ \rho^{-1} = Id[I(\theta)]$ (since $I(\theta) = D(\rho)$). Now, for every $x \in D(\sigma)$, $\sigma(x) = \rho(\theta(x))$, and so $\rho^{-1}(\sigma(x)) = \rho \circ \rho^{-1}(\theta(x)) = \theta(x)$, with the result that $\theta = \sigma \circ \rho^{-1}[D(\sigma)]$. But since $D(\rho^{-1}) \cap Var(S) = \emptyset$, then also $\theta = \sigma \circ \rho^{-1}[Var(S)]$. Now suppose $\theta' \in U(S)$, so that $\theta' = \theta \circ \eta$ for some η. Then $\theta' = \sigma \circ \rho^{-1} \circ \eta[Var(S)]$ and finally $\sigma \leq \theta'[Var(S)]$. $\square$

The following corollary will be used in a later result.

Corollary 3.3.12 If σ is a $mgu(S)[V]$ for some S and some V, then for every $\theta' \in U(S)$ we have $\sigma \leq \theta'[Var(S) \cup V]$.

Proof. By examining the details of the previous proof, we see that in fact $\theta = \sigma \circ \rho^{-1}[Var(S) \cup V]$, since $D(\rho^{-1}) \cap V = \emptyset$, and so $\theta' = \sigma \circ \rho^{-1} \circ \eta[Var(S) \cup V]$ and finally $\sigma \leq \theta'[Var(S) \cup V]$. $\square$

3.4 Equational Logic

In this section we review the basic notions of the model and proof theory of equational logic. First we recall some simple concepts regarding relations on terms.

Definition 3.4.1 Let $\longrightarrow$ be a binary relation on a set A, that is, $\longrightarrow \subseteq A \times A$. The *converse* (or *inverse*) of the relation $\longrightarrow$ is the relation denoted as $\longrightarrow^{-1}$ or $\longleftarrow$, defined such that $u \longleftarrow v$ iff $v \longrightarrow u$. The symmetric closure of $\longrightarrow$, denoted by $\longleftrightarrow$, is the relation $\longrightarrow \cup \longleftarrow$. The transitive closure, the reflexive and transitive closure, and the reflexive, symmetric, and transitive closure of $\longrightarrow$ are denoted respectively by $\overset{+}{\longrightarrow}$, $\overset{*}{\longrightarrow}$, and $\overset{*}{\longleftrightarrow}$. The *n*-fold composition of $\longrightarrow$ is denoted by $\overset{n}{\longrightarrow}$.

Definition 3.4.2 Let $\longrightarrow \subseteq T_\Sigma(X) \times T_\Sigma(X)$ be a binary relation on terms. The relation $\longrightarrow$ is *monotonic* iff for any three terms s, t, and u, for any $\alpha \in Dom(u)$, if $s \longrightarrow t$, then $u[\alpha \leftarrow s] \longrightarrow u[\alpha \leftarrow t]$ (sometimes a monotonic relation is called a *precongruence*). The relation $\longrightarrow$ is *stable* (under substitution) if $s \longrightarrow t$ implies that $\sigma(s) \longrightarrow \sigma(t)$ for every substitution σ.

Definition 3.4.3 Let $E \subseteq T_\Sigma(X) \times T_\Sigma(X)$ be a set of equations. We define the relation $\longleftrightarrow_E$ over $T_\Sigma(X)$ as the smallest symmetric, stable, and monotonic relation that contains E. This relation is defined explicitly as follows: Given any two terms $t_1, t_2 \in T_\Sigma(X)$, we have $t_1 \longleftrightarrow_E t_2$ iff there is some variant[5] $s \doteq t$ of an equation in $E \cup E^{-1}$, some tree address α in t_1, and some substitution σ, such that

$$t_1/\alpha = \sigma(s), \quad \text{and} \quad t_2 = t_1[\alpha \leftarrow \sigma(t)].$$

(Thus, σ is a matching substitution of s onto t_1/α.) Note that the equation $s \approx t$ is used as a two-way rewrite rule (that is, non-oriented). When $t_1 \longleftrightarrow_E t_2$, we say that we have an *equality step*. When we want to fully specify an equality step, we use the notation

$$t_1 \longleftrightarrow_{[\alpha, s \doteq t, \sigma]} t_2$$

---

[5] In what follows we shall assume that before an equation is used it has been renamed apart from all variables in current use. This is essential to prevent clashes among the variables. Thus we shall always state that a *variant* of an equation is being used.

(where some of the parameters may be omitted).

It is well known that the reflexive and transitive closure $\overset{*}{\longleftrightarrow}_E$ of $\longleftrightarrow_E$ is the smallest stable congruence on $T_\Sigma(X)$ containing E. This relation is the central object of study in the proof theory of equational logic, and is called E-congruence. Note that from our previous definitions, for any two terms u and v and set of equations E, we have $u \overset{*}{\longleftrightarrow}_E v$ iff there exists some sequence of equality steps

$$u = u_0 \longleftrightarrow_{[\alpha_1, l_1 \doteq r_1, \rho_1]} u_1 \longleftrightarrow_{[\alpha_2, l_2 \doteq r_2, \rho_2]} u_2 \cdots \longleftrightarrow_{[\alpha_n, l_n \doteq r_n, \rho_n]} u_n = v,$$

for $n \geq 0$, where the sequence of equations $l_i \doteq r_i$ consists of variants of equations from $E \cup E^{-1}$.

The decision problem associated with E-congruence, i.e., for a given E and two terms u and v, whether $u \overset{*}{\longleftrightarrow}_E v$, is called the *word problem* for E. That this is not a decidable problem is shown by our next result.

Theorem 3.4.4 The relation $\overset{*}{\longleftrightarrow}_E$ is in general only semi-decidable.

Proof. It should be obvious that it is decidable whether a given sequence of rewrite steps proves that $u \overset{*}{\longleftrightarrow}_E v$, and by the previous results, if $u \overset{*}{\longleftrightarrow}_E v$ for some set E, then there must exist a finite equational proof of this fact. Thus, by dovetailing the enumeration of all possible proofs, this sequence must eventually be discovered. But by picking E to contain the axioms for monoids, we see that $\overset{*}{\longleftrightarrow}_E$ can never be decidable in general, since the word problem for monoids is only semi-decidable (see [108]). $\square$

(For different proof of this result, see [36], where it is shown that any Turing Machine can be represented by an equational theory with only two equations.)

We now sketch the model theory of equational logic, where the language under consideration has equality and a set of function and constant symbols Σ, but no predicate symbols, and where the only sentences allowed are atomic, i.e., equations over $T_\Sigma(X)$. The only theories considered consist of sets of equational axioms, where all variables are implicitly universally quantified. Validity in this formal system is defined as follows.

Definition 3.4.5 Let $\mathcal{A}$ be a Σ-algebra. Since an equation is implicitly universally quantified, i.e., $l \doteq r$ is interpreted as the atomic formula $\forall x_1 \ldots x_n.(l \doteq r)$, where $\{x_1, \ldots, x_n\} = Var(l, r)$, we must define the semantics of equations with respect to *assignments* $\varphi : X \to A$. First, the meaning of a term t in the algebra $\mathcal{A}$ with respect to an assignment φ, denoted $[\![t]\!]_{\mathcal{A}}^{\varphi}$, is defined recursively on the structure of t as follows:

$$[\![x]\!]_{\mathcal{A}}^{\varphi} = \varphi(x), \quad \text{for a variable } x;$$
$$[\![c]\!]_{\mathcal{A}}^{\varphi} = c_{\mathcal{A}}, \quad \text{for a constant } c;$$
$$[\![f(t_1, \ldots, t_n)]\!]_{\mathcal{A}}^{\varphi} = f_{\mathcal{A}}([\![t_1]\!]_{\mathcal{A}}^{\varphi}, \ldots, [\![t_n]\!]_{\mathcal{A}}^{\varphi}).$$

(When the assignment φ is not significant, we omit it.) Using this definition, we say that

$$\mathcal{A} \models (l \doteq r)^{\varphi},$$

that is, $\mathcal{A}$ satisfies $l \doteq r$ with the assignment φ, iff we have

$$[\![l]\!]_{\mathcal{A}}^{\varphi} = [\![r]\!]_{\mathcal{A}}^{\varphi}.$$

Thus, the algebra $\mathcal{A}$ satisfies the equation $l \doteq r$, or is a *model* of $l \doteq r$, denoted

$$\mathcal{A} \models (l \doteq r),$$

iff for every assignment φ we have $\mathcal{A} \models (l \doteq r)^{\varphi}$. An algebra $\mathcal{A}$ is a model of an equational theory E if it is a model of every equation in E. The *variety* of a theory E is the class of all models of E. Finally, we say that a set of equations E *logically implies* an equation $l \doteq r$, denoted $E \models (l \doteq r)$, iff any model of E is also a model of $l \doteq r$.[6]

We conclude with the major result of equational logic, which asserts the soundness and completeness of the syntactic notion of E-congruence with respect to the model-theoretic semantics just presented. This result assures us that two terms are semantically equal modulo an equational theory E if and only if they can be proved congruent, by syntactically substituting equals for equals (i.e., terms equal under the theory E), in a finite number of steps.

Theorem 3.4.6 (Birkhoff) For any set of equations E and terms s and t, $E \models s \doteq t$ iff $s \xleftrightarrow{\ *\ }_E t$.

Before proceeding to the proof, we first define a "standard model" for a set of equations.

Definition 3.4.7 For any E, let T/E be the structure with universe $T_{\Sigma}(X)/\xleftrightarrow{\ *\ }_E$ (i.e., the set of congruence classes of $T_{\Sigma}(X)$ such that $[s] = [t]$ iff $s \xleftrightarrow{\ *\ }_E t$) and such that, for every $f \in \Sigma_n$, for $n \geq 0$,

$$f_{T/E}([t_1], \ldots, [t_n]) = [f(t_1, \ldots, t_n)]$$

---

[6] We remark here that we have not considered negations of equations or compound sentences made up of equations, since this will not be needed in what follows. The reader interested in a fuller treatment may consult [49].

(in particular, $c_{T/E} = [c]$).

It is easy to see that this interpretation of the function symbols enforces an interpretation of ground terms such that for any $t \in T_\Sigma$, the meaning the T/E assigns to t is simply $[t]$. Recall that for any function $\sigma : X \to T_\Sigma(X)$, there is always a unique homomorphic extension of σ to $T_\Sigma(X)$, namely, $\hat{\sigma} : T_\Sigma(X) \to T_\Sigma(X)$. In the sequel we shall identify $\hat{\sigma}$ and σ. Before we prove that T/E is indeed a model of the set E, we need the following technical lemma.

Lemma 3.4.8 For any E, let $\varphi : X \to T_\Sigma(X)/\overset{*}{\longleftrightarrow}_E$ be some assignment of the variables into the universe of T/E, and let $\sigma : X \to T_\Sigma(X)$ be any function such that $\sigma(x) \in \varphi(x)$ for every $x \in X$ (i.e., σ picks "representatives" for the classes in the range of φ). Then for any term t, $[\![t]\!]^\varphi_{T/E} = [\sigma(t)]$.

Proof. (By induction on $|t|$.) If $t = c \in \Sigma_0$, then $[\![c]\!]^\varphi_{T/E} = c_{T/E} = [c] = [\sigma(c)]$; and if $t = x \in X$, then $[\![x]\!]^\varphi_{T/E} = \varphi(x) = [\sigma(x)]$, since $\sigma(x) \in \varphi(x)$. Now suppose $t = f(t_1, \ldots, t_n)$ for $n > 1$. Then

$$
\begin{aligned}
[\![f(t_1, \ldots, t_n)]\!]^\varphi_{T/E} &= f_{T/E}([\![t_1]\!]^\varphi_{T/E}, \ldots, [\![t_n]\!]^\varphi_{T/E}) \\
&= f_{T/E}([\sigma(t_1)], \ldots, [\sigma(t_n)]) \\
&= [f(\sigma(t_1), \ldots, \sigma(t_n))] \\
&= [\sigma(f(t_1, \ldots, t_n))],
\end{aligned}
$$

where in the second step we applied the induction hypothesis. $\square$

Lemma 3.4.9 For any set of equations E, $T/E \models E$.

Proof. Suppose $l \doteq r \in E$; we must prove that $[\![l]\!]^\varphi_{T/E} = [\![r]\!]^\varphi_{T/E}$ for any assignment φ (which interpretes the universally quantified variables in $l \doteq r$). But then there must exist some σ as specified in the previous lemma (for example let $\sigma(x)$ be the least member of $\varphi(x)$ wrt some total ordering on $T_\Sigma(X)$). But then $[\sigma(l)] = [\sigma(r)]$, since $\sigma(l) \overset{*}{\longleftrightarrow}_E \sigma(r)$, and so by the previous lemma we have

$$
[\![l]\!]^\varphi_{T/E} = [\sigma(l)] = [\sigma(r)] = [\![r]\!]^\varphi_{T/E}.
$$

$\square$

Proof of Theorem 3.4.6.

For the *only if* direction (completeness), since T/E is a model of E, then $T/E \models s \doteq t$, i.e., $[\![l]\!]^\varphi_{T/E} = [\![r]\!]^\varphi_{T/E}$ for any assignment φ. Thus let

φ be the particular assignment which maps each x to $[x]$, and let σ be the identity on X. Clearly $\sigma(x) \in \varphi(x)$ for any x, and so by Lemma 3.4.8,

$$[s] = [\sigma(s)] = [\![l]\!]^{\varphi}_{T/E} = [\![r]\!]^{\varphi}_{T/E} = [\sigma(t)] = [t],$$

which means that $s \xleftrightarrow{*}_E t$.

For the *if* direction (soundness), first we prove that $E \models \rho(l) \doteq \rho(r)$ for any substitution ρ and $l \doteq r \in E$, by showing that $[\![\rho(l)]\!]^{\varphi}_{\mathcal{A}} = [\![\rho(r)]\!]^{\varphi}_{\mathcal{A}}$ for any model $\mathcal{A}$ of E and any assignment $\varphi : X \to A$. For any such $\mathcal{A}$ and φ, let $\theta : X \to A$ be the assignment such that for every x, $\theta(x) = [\![\rho(x)]\!]^{\varphi}_{\mathcal{A}}$ (or, equivalently, $\theta = \rho \circ \hat{\varphi}$, where $\hat{\varphi} : T_{\Sigma}(X) \to A$ is the homomorphic extension of φ).

We claim that $[\![\rho(t)]\!]^{\varphi}_{\mathcal{A}} = [\![t]\!]^{\theta}_{\mathcal{A}}$ for any $t \in T_{\Sigma}(X)$, and proceed by induction on $|t|$. If $t = c \in \Sigma_0$, then $[\![c]\!]^{\varphi}_{\mathcal{A}} = c_{\mathcal{A}} = [c] = [\![c]\!]^{\theta}_{\mathcal{A}}$; and if $t = x \in X$, then $[\![x]\!]^{\varphi}_{\mathcal{A}} = \varphi(x) = [\![x]\!]^{\theta}_{\mathcal{A}}$. Now suppose that $t = f(t_1, \ldots, t_n)$ for $n > 0$. Then

$$\begin{aligned}
[\![\rho(f(t_1, \ldots, t_n))]\!]^{\varphi}_{\mathcal{A}} &= [\![f(\rho(t_1), \ldots, \rho(t_n))]\!]^{\varphi}_{\mathcal{A}} \\
&= f_{\mathcal{A}}([\![\rho(t_1)]\!]^{\varphi}_{\mathcal{A}}, \ldots, [\![\rho(t_n)]\!]^{\varphi}_{\mathcal{A}}) \\
&= f_{\mathcal{A}}([\![t_1]\!]^{\theta}_{\mathcal{A}}, \ldots, [\![t_n]\!]^{\theta}_{\mathcal{A}}) \\
&= [\![f(t_1, \ldots, t_n)]\!]^{\theta}_{\mathcal{A}}
\end{aligned}$$

where in the second line we applied the induction hypothesis. This concludes the proof of the claim.

Now because $\mathcal{A} \models l \doteq r$, we have $[\![l]\!]^{\theta}_{\mathcal{A}} = [\![r]\!]^{\theta}_{\mathcal{A}}$, and so

$$[\![\rho(l)]\!]^{\varphi}_{\mathcal{A}} = [\![l]\!]^{\theta}_{\mathcal{A}} = [\![r]\!]^{\theta}_{\mathcal{A}} = [\![\rho(r)]\!]^{\varphi}_{\mathcal{A}},$$

and since φ was arbitrary, we conclude that $E \models \rho(l) \doteq \rho(r)$.

Next we show that if $E \models s \doteq t$, then $E \models u[s] \doteq u[t]$ for any context $u[]$, by induction on $|u|$. For the base case, if $|u| = 1$, then $u[s] = s$ and $u[t] = t$ and the result is trivial. Now suppose $|u| > 1$. Again, if $u[s] = s$ and $u[t] = t$ the result is trivial. Otherwise, if $u = f(u_1, \ldots, u_n)$, then $u[s] = f(u_1, \ldots, u_i[s], \ldots, u_n)$ and $u[t] = f(u_1, \ldots, u_i[t], \ldots, u_n)$ for some i, $1 \leq i \leq n$, and by hypothesis $E \models u_i[s] \doteq u_i[t]$, and then (since the rest of the context is identical) clearly

$$E \models f(u_1, \ldots, u_i[s], \ldots, u_n) \doteq f(u_1, \ldots, u_i[t], \ldots, u_n).$$

Finally, if $s \xleftrightarrow{*}_E t$, then we prove the soundness of this rewrite proof by induction on the number of rewrite steps n between s and t. If $n = 0$, then $s = t$ and the result is trivial. If $n > 0$, then either $s \longrightarrow_E s_1 \xleftrightarrow{*}_E t$, or $s \longleftarrow_E s_1 \xleftrightarrow{*}_E t$, where $E \models s_1 \doteq t$ by the induction hypothesis. We prove the first case, the second being similar. By the definition of $\longrightarrow_E$, $s = s[\rho(l)]$ and $s_1 = s[\rho(r)]$ for some $l \doteq r \in E$ and some substitution ρ. But then from the previous two paragraphs, we know that $E \models s[\rho(l)] \doteq s[\rho(r)]$, since clearly $E \models l \doteq r$, and we conclude that $[\![s]\!]_{\mathcal{A}}^{\varphi} = [\![s_1]\!]_{\mathcal{A}}^{\varphi} = [\![t]\!]_{\mathcal{A}}^{\varphi}$ for any model $\mathcal{A}$ of E and any assignment $\varphi : X \to A$, and so $E \models s \doteq t$. $\square$

3.5 Term Rewriting

One of the major problems with theorem proving in equational logic is that, even if a particular theory E has a decidable word problem, the search space for equational proofs is homogeneous and circular, so that a naive proof method has little choice but to enumerate proofs in some complete, brute force fashion. This means that even if the theory E has a decidable word problem, unless some ad-hoc method can be found, it may not be obvious how to design a decision procedure. The technique which has had the most success in the face of this combinatorial explosion is to attempt to impose a well-founded ordering on the set of terms, and to *orient* equations into rewrite rules which are always used in the same direction, thereby reducing the size of the terms being rewritten at each stage of the proof. If in addition the oriented set of equations forms a confluent relation, then we can find if two terms are congruent by simply reducing each to a (unique) normal form under the relation. If the set of oriented equations is not confluent, then sometimes it is possible to *complete* the set of equations by adding equational consequences in such a way that the result *is* confluent. This scheme has many advantages, although it can not be used in every case. In this section and the next, we review the major results in this area, and show how in some cases, by using the notion of terminating and confluent sets of rewrite rules, we can transform a set of equations which has a decidable word problem into a set of rules for which there is a trivial algorithm to solve the word problem.

Definition 3.5.1 When an equation $s \doteq t \in E$ is used only in one direction (from left to right), we call it a *rule*. The *reduction relation* $\longrightarrow_E$ is the smallest stable and monotonic relation that contains E. We can define $t_1 \longrightarrow_E t_2$ explicitly as in definition 3.4.3, the only difference being that $s \doteq t$ is a variant of an equation in E (and not in $E \cup E^{-1}$). When $t_1 \longrightarrow_E t_2$, we say that t_1 *rewrites* to t_2, or that we have a *rewrite step*. When $Var(r) \subseteq Var(l)$, then a rule is called a *rewrite rule* and denoted by

$l \twoheadrightarrow r$; a set of such rules is called a *rewrite system*.[7]

3.5.1 Termination Orderings

We now proceed to discuss the conditions under which a rewrite system can be proved to always terminate, first reviewing the basic notions of orderings we shall need.

Definition 3.5.2 A *preorder* (or *quasi-order*) $\preceq$ on a set A is a binary relation $\preceq \subseteq A \times A$ that is reflexive and transitive. A *partial order* $\preceq$ on a set A is a preorder that is also antisymmetric. The converse of a preorder (or partial order) $\preceq$ is denoted as $\succeq$. A *strict partial order* (or *strict order*) $\prec$ on a set A is a transitive and irreflexive relation. A pair $(A, \preceq)$ where $\preceq$ is a partial order on A is called a *poset*. Following [104] and [36], we also use the name 'poset' for a pair $(A, \prec)$ where $\prec$ is a strict order on A. Given a preorder (or partial order) $\preceq$ on a set A, the strict order $\prec$ associated with $\preceq$ is defined such that $s \prec t$ iff $s \preceq t$ and $t \not\preceq s$. Conversely, given a strict order $\prec$, the partial order $\preceq$ associated with $\prec$ is defined such that $s \preceq t$ iff $s \prec t$ or $s = t$. The converse of a strict order $\prec$ is denoted by $\succ$. Note that $(A, \succ)$ and $(A, \succeq)$ are posets whenever $(A, \prec)$ and $(A, \preceq)$ are posets. A partial order $\preceq$ (respectively a strict order $\prec$) on a set A is said to be *total* iff for any two *distinct* elements a and b of A, we have either $a \preceq b$ or $b \preceq a$ (respectively $a \prec b$ or $b \prec a$).[8]

Definition 3.5.3 A relation $\succ$ on a set A is *Noetherian* or *well-founded* iff there are no infinite sequences $a_0, \ldots, a_n, a_{n+1}, \ldots$ of elements in A such that $a_n \succ a_{n+1}$ for all $n \geq 0$. Given a preorder (or partial order) $\preceq$, we say that $\preceq$ is well-founded iff $\succ$ is well-founded. A set of rewrite rules R is called noetherian if the relation $\longrightarrow_R$ is noetherian.

Remark: We warn our readers that this is not the usual way of defining a well-founded relation in set theory, as for example in Levy [104]. In set theory, the condition is stated in the form $a_{n+1} \prec a_n$ for all $n \geq 0$, where $\prec = \succ^{-1}$. It is the dual of the condition we have used, but since $\prec = \succ^{-1}$, the two definitions are equivalent. When using well-founded relations in the context of rewriting systems, since it is customary to give rewrite sequences

---

[7] The motivation for this restriction is that whenever there is some variable $x \in Var(r) - Var(l)$ then there can never exist a stable simplification ordering for this rule, since we may apply the substitution $[l/x]$ to the rule, for example. Such pathological cases are eliminated by the restriction; cf. the problems which arise with global variables in functional programming languages.

[8] Clearly if $\preceq$ is a partial order, then this holds for *any* two elements of A, distinct or not.

from left to right, we are usually interested in the reduction relation $\longrightarrow$ and the fact that there are no infinite sequences $a_0, \ldots, a_n, a_{n+1}, \ldots$ such that $a_n \longrightarrow a_{n+1}$ for all $n \geq 0$. Thus, following other authors, including Dershowitz, we adopt the dual of the standard set theoretic definition. For the same reason, in the next two definitions we use posets of the form $(A, \succ)$ rather than of the form $(A, \prec)$.

Definition 3.5.4 For each positive integer n and n posets $(S_i, \succ_i)$, for $1 \leq i \leq n$, we may define the *lexicographic order* $\succ^{lex}$ on the set $S_1 \times \ldots \times S_n$ as follows. Let $a_1, \ldots, a_n$ and $b_1, \ldots, b_n$ be members of $S_1 \times \ldots \times S_n$. Then

$$a_1, \ldots, a_n \quad \succ^{lex} \quad b_1, \ldots, b_n$$

if and only if there exists some $1 \leq i \leq n$ such that $a_i \succ_i b_i$, and for all $1 \leq j < i$, $a_j = b_j$. Unless stated otherwise, we will normally assume that all the n posets are identical.

Definition 3.5.5 Let $(S, \succ)$ be a poset, let M be some finite multiset of objects from S, and finally let $n, n'_1, \ldots, n'_k \in S$. Define the transformation (relation) $\Rightarrow_m$ on finite multisets as

$$M \cup \{n\} \quad \Rightarrow_m \quad M \cup \{n'_1, \ldots, n'_k\},$$

where $k \geq 0$ and $n \succ n'_i$ for all i, $1 \leq i \leq k$. Then the multiset ordering $\gg$ is simply the transitive closure $\overset{+}{\Rightarrow}_m$. In other words, $N \gg N'$ iff N' is produced from a finite multiset N by removing one or more elements and replacing them with any finite number of elements, each of which is strictly smaller than at least one element removed.

For a given strict order $\succ_{name}$ we denote the corresponding multiset extension by $\gg_{name}$ and the lexicographic extension by $\succ^{lex}_{name}$. We may extend these to partial orders as shown in Definition 3.5.2.

It is easy to show that for any poset $(S, \succ)$ we have associated posets $(M, \gg)$ (where M is the set of all finite multisets of members of S) and $(S^n, \succ^{lex})$ for $n > 0$. Furthermore $\succ$ is total (respectively, well-founded) iff $\succ^{lex}$ (for any n) is total (respectively, well-founded) iff $\gg$ is total (respectively, well-founded).

We are interested of course in using orderings on terms to prove termination of rewrite systems, and, more generally, to do inductive proofs. The most general of these orderings are based on the notion of syntactic simplification.

Definition 3.5.6 A strict ordering $\prec$ has the *subterm property* if $s \prec f(\ldots, s, \ldots)$ for every term $f(\ldots, s, \ldots)$. A *simplification ordering* $\prec$ is a strict ordering that is monotonic and has the subterm property (since we are considering symbols having a fixed rank, the deletion property is superfluous, as noted in Dershowitz [36]). A *reduction ordering* $\prec$ is a strict ordering that is monotonic, stable, and such that $\succ$ is well-founded. With a slight abuse of language, we will also say that the converse $\succ$ of a strict ordering $\prec$ is a simplification ordering (or a reduction ordering).

The importance of these orderings is shown by this next fundamental result, from [35].

Lemma 3.5.7 A set of rules R is noetherian if and only if there exists a reduction ordering $\succ$ on $T_\Sigma(X)$ such that for every $l \overset{.}{\to} r \in R,\ l \succ r$.

Unfortunately, it is undecidable in general if an arbitrary system R is noetherian [75], since it is possible to encode Turing machines using rewrite rules [36], and this would imply the decidability of the halting problem.

The most powerful forms of reduction orderings are based on the relative syntactic simplicity of two terms, i.e., on the notion of a simplification ordering. Although there are many types of simplification orderings, one of the most elegant and useful is the *recursive path ordering with status*, and since we shall not make detailed use of term orderings in this thesis, we shall content ourselves with presenting just this one ordering.

Let a *precedence ordering* be an irreflexive partial order $\succ$ on Σ. The *multiset path ordering* (also called the *recursive path ordering*) and *lexicographic path ordering* are two well–known methods for extending a precedence ordering to terms; a generalization encompassing both may be given as follows.[9]

Definition 3.5.8 Let us assume every function symbol in Σ is assigned a *status* from the set $\{mul, lex\}$. The *recursive path ordering with status* is defined as follows. For any two $s, t \in T(\Sigma)$, $s =_{rpos} t$ iff s and t are identical up to permutation of immediate subterms under function symbols with *mul* status; and

$$s = f(s_1, \ldots, s_n) \succ_{rpos} g(t_1, \ldots, t_m) = t$$

iff

(i) $f \succ g$ and $s \succ_{rpos} t_i$ for *all* i, $1 \le i \le m$; or
(ii) $s_i \succeq_{rpos} t$ for *some* i, $1 \le i \le n$; or

---

[9] In [36], various orderings are defined in terms of of a quasi-order on Σ; we shall not use this increased generality in this note, and so we give a simpler definition.

(iii) $f = g$ has *mul* status and $\{s_1, \ldots, s_n\} \succ^{mul}_{rpos} \{t_1, \ldots, t_m\}$; or
(iv) $f = g$ has *lex* status and $(s_1, \ldots, s_n) \succ^{lex}_{rpos} (t_1, \ldots, t_n)$ and
      $s \succ_{rpos} t_i$ for *all* i, $1 \leq i \leq m$.

(For the purposes of this monograph, the ordering can be extended to $T(\Sigma, X)$ by considering the variables to be constants incomparable with any other symbols.) The ordering obtained when all function symbols have *mul* status is simply the *multiset path ordering* [36], and when all have *lex* status, the *lexicographic path ordering*, due to [84]. It is well–known that $\succ_{rpos}$ is well–founded when $\succ$ is, for any status, and that when $\succ$ is total, then so is the *lpo*, although if any function symbols have *mul* status, then $\succeq_{rpos}$ is at most a preorder.

For our purposes, the most important thing about this definition is that for any signature, there exists a reduction ordering total on ground terms, which is also a simplification ordering. (This result will be necessary in the proof of the completeness of our set of transformations $\mathcal{T}$ given in Chapter §6.) However, note that this ordering may not be total on $T_\Sigma(X)$. This is a major problem with term orderings: in order to preserve stability under substitution, they must treat variables as incomparable symbols. Thus equations such as commutative axioms (e.g. $f(x, y) \doteq f(y, x)$) can not be oriented. A more general discussion of term orderings may be found in [35, 36, 77].

3.5.2 Confluence

For the rest of this section, we discuss the confluence of rewrite systems and an important result which provides a sufficient condition for a noetherian system to be confluent.

Definition 3.5.9 A system R is *confluent* if for any terms t, t_1, and t_2, whenever $t_1 \overset{*}{\longleftarrow}_R t \overset{*}{\longrightarrow}_R t_2$, there exists a w such that $t_1 \overset{*}{\longrightarrow}_R w \overset{*}{\longleftarrow}_R t_2$. We shall call such a w the *confluence term* for t_1 and t_2, and shall write $t_1 \downarrow t_2$ if there exists a confluence term for t_1 and t_2.

The following definition and lemma provide an alternate characterization of confluence.

Definition 3.5.10 A system R is *Church-Rosser* if for any terms t_1 and t_2, $t_1 \overset{*}{\longleftrightarrow}_R t_2$ implies that $t_1 \downarrow t_2$.

Lemma 3.5.11 A system R of rules is confluent iff it is Church-Rosser.

Proof. The *if* part is trivial, since $u \overset{*}{\longleftarrow}_R t \overset{*}{\longrightarrow}_R v$ implies that $u \overset{*}{\longleftrightarrow}_R v$, and hence that $u \downarrow v$. The *only if* part proceeds by induction on the number of rewrite steps in $\overset{*}{\longleftrightarrow}_R$. The base case if trivial. Now suppose the result holds for all sequences of less than n rewrite steps, and let

$$s = s_0 \longleftrightarrow_R s_1 \longleftrightarrow_R \ldots s_{n-1} \longleftrightarrow_R s_n = t.$$

By the hypothesis, there is some w such that $s \overset{*}{\longrightarrow}_R w \overset{*}{\longleftarrow}_R s_{n-1}$ and some v such that $s_{n-1} \overset{*}{\longrightarrow}_R v \overset{*}{\longleftarrow}_R t$, regardless of the direction of the last rewrite step. But then $w \overset{*}{\longleftarrow}_R s_{n-1} \overset{*}{\longrightarrow}_R v$, and by confluence, there must exist some u such that $s \overset{*}{\longrightarrow}_R w \overset{*}{\longrightarrow}_R u \overset{*}{\longleftarrow}_R v \overset{*}{\longleftarrow}_R t$, as illustrated by:

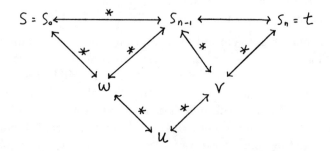

$\square$

A rewrite system R is termed *canonical* if it is noetherian and confluent. The significance of canonical systems of rewrite rules is that they have a decidable word problem, as shown by the next two results.

Definition 3.5.12 Let R be a set of rewrite rules. A term t is in *normal form with respect to R* or in *R-normal form* if for no t' does $t \longrightarrow_R t'$. If R is available from context we shall simply say that t is in *normal form*.

Lemma 3.5.13 Let R be a canonical system of rewrite rules. Then any term t has a unique normal form with respect to R.

Proof. Suppose $u \overset{*}{\longleftarrow}_R t \overset{*}{\longrightarrow}_R v$, with u and v in normal form. Then by the confluence of R, there must exist a term w such that $u \overset{*}{\longrightarrow}_R w \overset{*}{\longleftarrow}_R v$. Since u and v are in normal form, this can only happen if $u = w = v$. $\square$

The unique normal form of a term t will be denoted by $t{\downarrow}$.

Theorem 3.5.14 Let R be a canonical set of rewrite rules and s and t be two terms. Then $s \overset{*}{\longleftrightarrow}_R t$ iff $s{\downarrow} = t{\downarrow}$.

Proof. The *if* part must hold since $s \overset{*}{\longrightarrow}_R s{\downarrow} = t{\downarrow} \overset{*}{\longleftarrow}_R t$. To prove *only if*, we observe that by Lemma 3.5.11, R is Church-Rosser, so there must exist some term w such that $s \xrightarrow{*}_R w \xleftarrow{*}_R t$. But then by the previous lemma, $w{\downarrow}$ is the unique normal form of w, s, and t. $\square$

This theorem gives us a simple decision procedure for the word problem in canonical theories: we simply reduce both terms to their unique normal forms and compare. Confluence is undecidable in general [77], but, in fact, if we can prove that the system has the termination property, it is possible to localize the test for confluence to single applications of rewrite rules. This is the key result which will allow us to define the completion of a set of equations in the next section.

Definition 3.5.15 A system R is *locally confluent* iff for terms t, t_1, and t_2, whenever $t_1 \xleftarrow{}_R t \xrightarrow{}_R t_2$, there exists a term w such that $t_1 \xrightarrow{*}_R w \xleftarrow{*}_R t_2$.

Theorem 3.5.16 A noetherian system R is confluent iff it is locally confluent.

Proof. The only difficulty is showing that a noetherian and locally confluent system is confluent, since the other direction is trivial. Thus suppose R is noetherian under $\succ$. We proceed by induction on the well-founded term ordering $\succ$ to show that for any term u, if $u_1 \xleftarrow{*}_R u \xrightarrow{*}_R u_2$, then there exists some term w such that $u_1 \xrightarrow{*}_R w \xleftarrow{*}_R u_2$. For the base case, if u is in normal form, then the result is trivial. Now assume the result holds for all terms u' such that $u \succ u'$, and suppose $u_1 \xleftarrow{n}_R u \xrightarrow{m}_R u_2$. If $n = 0$ then let $w = u_2$; if $m = 0$ let $w = u_1$. Otherwise, we have

$$u_1 \xleftarrow{n-1}_R u_1' \xleftarrow{}_R u \xrightarrow{}_R u_2' \xrightarrow{m-1}_R u_2,$$

where by local confluence there must exist some term v such that $u_1' \xrightarrow{*}_R v \xleftarrow{*}_R u_2'$, and by the induction hypothesis (applied to u_1'), there must exist a term v_1 such that $u_1 \xrightarrow{*}_R v_1 \xleftarrow{*}_R v$, and, finally, by the induction hypothesis (applied to u_2'), there must exist a term w such that $v_1 \xrightarrow{*}_R w \xleftarrow{*}_R u_2$, as illustrated by:

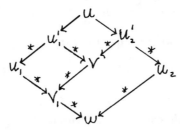

$\square$

This important result shows us that when considering a noetherian set of rewrite rules, we can localize the test for confluence, in the sense that the notion of local confluence centers on the application of two rewrite rules to an arbitrary term t. By considering the possible ways that two rules can interact in rewriting a single term, we show in our next theorem that a system can only fail to be locally confluent when a particular kind of overlap is possible between the matching of rules onto a term. In this way we can define the sufficient condition for local confluence which motivates the design of the completion procedure presented in the next section.

Definition 3.5.17 Suppose $t_2 \longleftarrow_{[\alpha_2, l_2 \dot\rightarrow r_2, \rho_2]} t \longrightarrow_{[\alpha_1, l_1 \dot\rightarrow r_1, \rho_1]} t_1$. Then we say there is a *critical overlap* of the two rules on t if the projections of the non-variable symbols in each of the terms l_1 and l_2 onto t intersect, that is, if $\{\alpha_1\beta \mid \beta \in NonVarDom(l_1)\} \cap \{\alpha_2\gamma \mid \gamma \in NonVarDom(l_2)\} \neq \emptyset$.

Before we present the theorem, we need one lemma which generalizes a proposition from [73].

Lemma 3.5.18 Let R be a set of rewrite rules and θ and θ' be substitutions with $D(\theta) = D(\theta')$ such that $\forall x \in D(\theta)$, $\theta(x) \overset{*}{\longrightarrow}_R \theta'(x)$. Then for any t, $\theta(t) \overset{*}{\longrightarrow}_R \theta'(t)$.

Proof. (By induction on $|t|$.)

Basis. $|t| = 0$. If $t \in D(\theta)$, then trivially $\theta(t) \overset{*}{\longrightarrow}_R \theta'(t)$; otherwise $\theta(t) = \theta'(t)$.

Induction. Assume for all terms of depth less than k, with $k > 0$. Then t of depth k must be in the form $f(t_1, \ldots, t_n)$, where each t_i has a depth no more than $k-1$ and $\theta(t_i) \overset{*}{\longrightarrow}_R \theta'(t_i)$ for $1 \leq i \leq n$. By concatenating these n disjoint rewrite sequences, we obtain

$$\theta(t) = f(\theta(t_1), \ldots, \theta(t_n)) \overset{*}{\longrightarrow}_R f(\theta'(t_1), \ldots, \theta'(t_n)) = \theta'(t).$$

$\square$

We now show that the existence of critical overlaps is a necessary condition for a system to be non-confluent.

Theorem 3.5.19 (Knuth-Bendix) Let t, t_1, and t_2 be terms and let

$$t_2 \longleftarrow_{[\alpha', l_2 \dot\rightarrow r_2, \rho_2]} t \longrightarrow_{[\alpha, l_1 \dot\rightarrow r_1, \rho_1]} t_1,$$

where $l_1 \dot\rightarrow r_1$ and $l_2 \dot\rightarrow r_2$ are variants of rules from some set R. Then either there exists a confluence term w such that $t_2 \overset{*}{\longrightarrow}_R w \overset{*}{\longleftarrow}_R t_1$ or there is a critical overlap of $l_1 \dot\rightarrow r_1$ and $l_2 \dot\rightarrow r_2$ on t.

Proof. As regards the locations of the addresses α and α' in t, there are two cases.

(A) α and α' are disjoint, in which case there must exist a w such that

$$t_2 \xrightarrow{}_{[\alpha,l_1 \dot{\rightarrow} r_1,\rho_1]} w \xleftarrow{}_{[\alpha',l_2 \dot{\rightarrow} r_2,\rho_2]} t_1,$$

in other words, the rewrite steps commute, which may be illustrated:

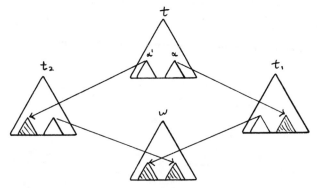

(B) One of α, α' is a prefix of the other. Without loss of generality, assume that $\alpha' = \alpha\beta$. Then $t/\alpha = \rho_1(l_1)$ and $t/\alpha\beta = \rho_1(l_1)/\beta = \rho_2(l_2)$, and we have two subcases: either (i) $\beta \notin NonVarDom(l_1)$, that is, $\rho_1(l_1)/\beta$ is a subterm of $\rho_1(x)$ for some $x \in Var(l_1)$, or (ii) $\beta \in NonVarDom(l_1)$. This latter case is a critical overlap, so if we can show the existence of a confluence term in case (i) we are done.

Let $\beta = \beta_1\beta_2$, where $l_1/\beta_1 = x$, so that $t/\alpha\beta = \rho_1(l_1)/\beta = \rho_1(x)/\beta_2 = \rho_2(l_2)$ as shown in the following figure:

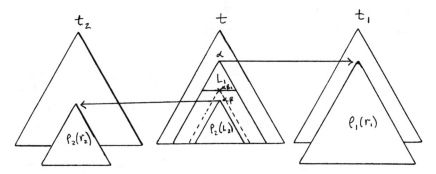

Now let the substitution ρ_1' be defined as

$$\rho_1'(y) = \begin{cases} \rho_1(y)[\beta_2 \leftarrow \rho_2(r_2)], & \text{if } y = x = l_1/\beta_1; \\ \rho_1(y) & \text{otherwise.} \end{cases}$$

We now show that $w = t[\alpha \leftarrow \rho_1'(r_1)]$ is a confluence term for t_1 and t_2. Clearly, $\rho_1(x) \longrightarrow_R \rho_1'(x)$, and so by Lemma 3.5.18, we have that

$\rho_1(r_1) \xrightarrow{*}_R \rho'_1(r_1)$. But then by monotonicity we have

$$t_1 = t[\alpha \leftarrow \rho_1(r_1)] \xrightarrow{*}_R t[\alpha \leftarrow \rho'_1(r_1)] = w.$$

To show that $t_2 \xrightarrow{*}_R w$, we observe that since $t/\alpha = \rho_1(l_1)$ and $t/\alpha\beta_1 = \rho_1(x)$, then $t_2/\alpha = \rho_1(l_1)[\beta_1 \leftarrow \rho'_1(x)]$. Now because the rewrite rules used were variants, $Var(\rho'_1(x)) \cap D(\rho_1) = \emptyset$ and therefore $\rho_1(\rho'_1(x)) = \rho'_1(x)$. Thus we have

$$\begin{aligned}
t_2/\alpha &= \rho_1(l_1)[\beta_1 \leftarrow \rho'_1(x)] \\
&= \rho_1(l_1[\beta_1 \leftarrow \rho'_1(x)]) \\
&\xrightarrow{*}_R \rho'_1(l_1[\beta_1 \leftarrow \rho'_1(x)]) \quad ; \text{ by Lemma 3.5.18} \\
&= \rho'_1(l_1).
\end{aligned}$$

But then by monotonicity, we must have

$$t_2 = t[\alpha \leftarrow t_2/\alpha] \xrightarrow{*}_R t[\alpha \leftarrow \rho'_1(l_1)] \xrightarrow{}_{[\alpha, l_1 \dot\to r_1, \rho'_1]} t[\alpha \leftarrow \rho'_1(r_1)] = w. \qquad \square$$

This result shows us that the test for local confluence may be restricted to critical overlaps. But the variant assumption for the use of rules allows us to sharpen this test so that we may confine our search for critical overlaps to an examination of the rules themselves.

Lemma 3.5.20 Suppose that $t_2 \xleftarrow{}_{[\alpha', l_2 \dot\to r_2, \rho_2]} t \xrightarrow{}_{[\alpha, l_1 \dot\to r_1, \rho_1]} t_1$, where (without loss of generality) $\alpha' = \alpha\beta$, and where $l_1 \dot\to r_1$ and $l_2 \dot\to r_2$ are two variants of (not necessarily distinct) rewrite rules. Then there exists a critical overlap of the two rules on t iff there exists a substitution $\sigma = mgu(l_1/\beta, l_2)$, where $\beta \in NonVarDom(l_1)$.

Proof. The *if* part of the proof is trivial, by taking $t = \sigma(l_1)$. For the other direction, since $\beta \in NonVarDom(l_1)$, we have $t/\alpha\beta = \rho_1(l_1)/\beta = \rho_1(l_1/\beta) = \rho_2(l_2)$. Now, by the variant assumption it must be the case that $Var(l_1) \cap Var(l_2) = \emptyset$, and so $D(\rho_1) \cap D(\rho_2) = \emptyset$, and thus $\rho_1 \cup \rho_2(l_1/\beta) = \rho_1 \cup \rho_2(l_2)$. Therefore there must exist a substitution $\sigma = mgu(l_1/\beta, l_2)$. $\qquad \square$

The point is that anytime there is a critical overlap, this must have been caused by a most general overlap between the two rules, independent of the term t in which the overlap occurs. This leads us to our next definition.

Definition 3.5.21 Let $l_1 \dot\to r_1$ and $l_2 \dot\to r_2$ be two variants of (not necessarily distinct) rewrite rules, and $\sigma = mgu(l_1/\beta, l_2)$, with β a non-variable position in l_1. Then the pair of terms p, q is called a *critical pair*, where $p = \sigma(l_1[\beta \leftarrow r_2])$ and $q = \sigma(r_1)$.

A critical pair is always a consequence of the set of rules, that is, $p \xleftrightarrow{*}_R q$, since $\sigma(l_1[\beta \leftarrow r_2]) \xleftarrow{}_R \sigma(l_1) \xrightarrow{}_R \sigma(r_1)$. We may think of a critical pair as a most general instance of the way in which two critically overlapping rules can rewrite a term.

Lemma 3.5.22 (Knuth-Bendix, Huet) Let $l_1 \xrightarrow{\cdot} r_1$ and $l_2 \xrightarrow{\cdot} r_2$ be variants from a set of rules R and let t, t_1, and t_2 be terms such that $t_2 \xleftarrow{}_{[\alpha\beta, l_2 \xrightarrow{\cdot} r_2, \rho_2]} t \xrightarrow{}_{[\alpha, l_1 \xrightarrow{\cdot} r_1, \rho_1]} t_1$, with $\beta \in NonVarDom(l_1)$, as shown by the following figure:

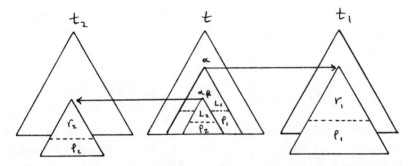

Then there exists a critical pair p, q for the two rules, and a substitution η such that $t_2/\alpha = \eta(p)$ and $t_1/\alpha = \eta(q)$.

Proof. By the previous lemma, there must exist a critical pair p, q and a substitution σ such that $p = \sigma(l_1[\beta \leftarrow r_2])$ and $q = \sigma(r_1)$. Now, since σ is a *mgu*, it must be the case that $\sigma \leq \rho_1 \cup \rho_2$, and so there must be some η such that $\rho_1 \cup \rho_2 = \sigma \circ \eta$. Also, by the variant assumption, $D(\rho_1) \cap Var(r_2) = \emptyset$ and $D(\rho_2) \cap Var(l_1) = \emptyset$. Therefore, we have

$$
\begin{aligned}
t_2/\alpha &= t/\alpha[\beta \leftarrow \rho_2(r_2)] \\
&= \rho_1(l_1)[\beta \leftarrow \rho_2(r_2)] \\
&= \rho_1 \cup \rho_2(l_1[\beta \leftarrow r_2]) \\
&= \eta(\sigma(l_1[\beta \leftarrow r_2])) \\
&= \eta(p).
\end{aligned}
$$

Finally, by the variant assumption, $D(\rho_2) \cap Var(r_1) = \emptyset$, and so

$$
\begin{aligned}
t_1/\alpha &= \rho_1(r_1) \\
&= \rho_1 \cup \rho_2(r_1) \\
&= \eta(\sigma(r_1)) \\
&= \eta(q).
\end{aligned}
$$

$\square$

We may now bring all these results together in presenting the major theorem of this section, which shows that the test for local confluence may be restricted to the critical pairs of a system of rewrite rules. Clearly, if the system is also noetherian, then this constitutes a test for confluence as well.

Theorem 3.5.23 (Knuth-Bendix, Huet) A set of rewrite rules R is locally confluent iff for every critical pair p, q formed from variants of rules in R, $p \downarrow q$.

Proof. ($\Rightarrow$) Suppose R is locally confluent and let p, q be a critical pair formed from variants $l_1 \dotrightarrow r_1$ and $l_2 \dotrightarrow r_2$ of rules from R. But then $p \longleftarrow_R \sigma(l_1) \longrightarrow_R q$, and by local confluence there must exist a term w such that $p \overset{*}{\longrightarrow}_R w \overset{*}{\longleftarrow}_R q$.

($\Leftarrow$) Suppose every critical pair has a confluence term, and let

$$t_2 \longleftarrow_{[\alpha', l_2 \dotrightarrow r_2, \rho_2]} t \longrightarrow_{[\alpha, l_1 \dotrightarrow r_2, \rho_1]} t_1.$$

By Theorem 3.5.19, we see that if there is no critical overlap, there must exist a confluence term for t_1 and t_2. If there is a critical overlap, i.e., $\beta \in NonVarDom(l_1)$, then by Lemma 3.5.22, there must exist a critical pair p, q and a substitution η such that $t_2/\alpha = \eta(p)$ and $t_1/\alpha = \eta(q)$. By the hypothesis, there must exist a term w such that $p \overset{*}{\longrightarrow}_R w \overset{*}{\longleftarrow}_R q$, and thus, by the stability of $\longrightarrow_R$, we have $\eta(p) \overset{*}{\longrightarrow}_R \eta(w) \overset{*}{\longleftarrow}_R \eta(q)$. But then $t_2 \overset{*}{\longrightarrow}_R t[\alpha \leftarrow \eta(w)] \overset{*}{\longleftarrow}_R t_1$, completing the proof. $\square$

The reader interested in further results concerning the confluence of term rewriting systems is referred to [73].

3.6 Completion of Equational Theories

The last result in our previous section showed us that a sufficient condition for the confluence of a finite noetherian system may be verified by checking, for each critical pair (of which there are only finitely many), if the normal forms of each term in the pair are equal.If this test fails, we may attempt to complete the system of rules by adding the critical pairs to the system as rules, if they can be oriented with respect to the termination ordering. Since each pair is a logical consequence of the original system, this is a conservative extension of the theory, and each of the original critical pairs is now trivially confluent. Unfortunately, new critical pairs may be created. By iterating this process, we may eventually succeed in finding a canonical system of rules R_n equivalent to the original system E, i.e., such that

$\overset{*}{\longleftrightarrow}_E \ = \ \overset{*}{\longleftrightarrow}_{R_n}$, and then we may use the decision procedure for R_n (implicit in Theorem 3.5.14) to solve the word problem in E. Of course, since the word problem for arbitrary E is not decidable in general, this method will not work in every case.

This is the basic method of the Knuth-Bendix Completion Procedure first presented in [94] and thereafter refined by many different researchers (for a good survey of the history of completion and its parallel development in polynomial ideal theory, with an excellent bibliography, see [24]). We shall follow [37] in presenting an abstract version of this procedure along the lines of [8]. The result will be a non-deterministic version of the Knuth-Bendix completion procedure.

Definition 3.6.1 (The set $\mathcal{KB}$) Let E be a set of equations, R be a set of rules (either possibly empty) and $\succ$ a reduction ordering on $T_\Sigma(X)$.

The first transformation simply removes trivial equations from E.

$$E \cup \{u \doteq u\}, R \implies E, R. \tag{1}$$

We may reduce equations in E by rules in R, so that if $u \longrightarrow_R w$,

$$E \cup \{u \doteq v\}, R \implies E \cup \{w \doteq v\}, R, \tag{2a}$$

and if $v \longrightarrow_R w$,

$$E \cup \{u \doteq v\}, R \implies E \cup \{u \doteq w\}, R. \tag{2b}$$

Furthermore, we may inter-reduce rules in R. If $v \longrightarrow_R w$, then

$$E, R \cup \{u \overset{.}{\rightarrow} v\} \implies E, R \cup \{u \doteq w\}, \tag{3}$$

and if $\triangleright$ is a well-founded ordering on rewrite rules and $u \longrightarrow_{[\alpha, l \overset{.}{\rightarrow} r, \rho]} w$, where $u \overset{.}{\rightarrow} v \triangleright l \overset{.}{\rightarrow} r$, then

$$E, R \cup \{u \overset{.}{\rightarrow} v\} \implies E \cup \{w \doteq v\}, R. \tag{4}$$

(Notice that in transformation (3), the new rule remains in R, since $u \succ v \succ w$, whereas in transformation (4), the new rule must be moved back to E, in case $w \not\succ v$.)

If $u \succ v$ in the reduction ordering, then

$$E \cup \{u \doteq v\}, R \implies E, R \cup \{u \overset{.}{\rightarrow} v\}. \tag{5}$$

The last transformation tries to make the system confluent by moving equational consequences into the equational theory. If $u \longleftarrow_R w \longrightarrow_R v$ then

$$E, R \implies E \cup \{u \doteq v\}, R. \tag{6}$$

Note in particular that u, v may be a critical pair of R. As in the transformations $\mathcal{ST}$, we assume that union on the left-hand side of each of these rules is a *disjoint union*.

The well-founded ordering $\rhd$ in rule (4) insures that the interreduction process is correct by forcing the rewrite rule $l \overset{.}{\to} r$ to be smaller in some sense than $u \overset{.}{\to} v$. The common practice is to require l to be a proper subterm of u or a proper substitution instance of u, but it can also be advantagous to consider the age of the rules. In no case can a rule be used to reduce its own left-hand side. (For details see [37] or [8].)

The soundness of the transformations is given by this next theorem, whose proof (although tedious) is not hard and is omitted.

Theorem 3.6.2 (Soundness of the set $\mathcal{KB}$) If $E, R \Longrightarrow_{KB} E', R'$ then $\overset{*}{\longleftrightarrow}_{E \cup R} = \overset{*}{\longleftrightarrow}_{E' \cup R'}$.

A *completion procedure* is a strategy for applying these transformations; specifically, it takes as input a finite set of equations E and a reduction ordering $\succ$ and generates a sequence of pairs

$$E, \emptyset \Longrightarrow_{KB} E_1, R_1 \Longrightarrow_{KB} E_2, R_2 \Longrightarrow_{KB} \cdots$$

It should be obvious that the systems $R_1, R_2, \ldots$ are all noetherian with respect to the reduction ordering $\succ$. The intent of the set $\mathcal{KB}$ is to find a finite sequence of transformations

$$E, \emptyset \Longrightarrow_{KB} E_1, R_1 \Longrightarrow_{KB} \ldots \emptyset, R_n,$$

where R_n is canonical (if possible). In particular, this will be true if R_n has has no new critical pairs (i.e. that have not already been generated). In this case, by Theorem 3.5.23, R_n is locally confluent and hence canonical under $\succ$. In this case we say that R is a *completion* for E. This shows that if a set of equations E is completed into a set of rules R, then $\overset{*}{\longleftrightarrow}_E = \overset{*}{\longleftrightarrow}_R$, i.e., the word problem in each is identical—except that there is a simple decision procedure for R.

If no transformation applies, but E is non-empty, the procedure is said to *fail* (this may happen, for instance, if there is some $l \overset{.}{=} r \in E$ such that $l \not\succ r$ and $r \not\succ l$). It is also possible that the procedure may run forever, even given certain fairness assumptions regarding the selection of equations and of critical pairs. In this case, we say that the result of a (possibly) infinite sequence

$$E, \emptyset \Longrightarrow_{KB} E_1, R_1 \Longrightarrow_{KB} \cdots$$

is E^∞, R^∞, where $E^\infty = \bigcup_{i \geq 0} \bigcap_{j \geq i} E_j$ is the set of *persisting equations* and $R^\infty = \bigcup_{i \geq 0} \bigcap_{j \geq i} R_j$ is the set of *persisting rewrite rules*. A sequence is called *successful* if $\bar{E}^\infty = \emptyset$ and R^∞ is canonical. A completion procedure (i.e. a strategy for applying transformations from $\mathcal{KB}$) is *correct* if $E^\infty = \emptyset$ implies that R^∞ is canonical (whether the sequence is infinite or not), and it is *complete* if every sequence under the strategy is successful. (For details see [37] and [8].)

In order to present the completeness result for $\mathcal{KB}$, we need the notion of a *fair* transformation sequence. Let $cp(R)$ be the set of all critical pairs formed from the rules in R, and let us say that a system R is *reduced* if for every $l \dashrightarrow r$ in R, r is in normal form with respect to R, and l is in normal form with respect to $R - \{l \dashrightarrow r\}$.

Definition 3.6.3   A $\mathcal{KB}$-transformation sequence is *fair* if $cp(R^\infty) \subseteq \bigcup_{i \geq 0} E_i$, R^∞ is reduced, and $E^\infty = \emptyset$.

The following result, from [76], gives the sense in which the set $\mathcal{KB}$ can be considered to be complete. (See [37] or [8] for a proof.)

Theorem 3.6.4   If a $\mathcal{KB}$-transformation sequence

$$E, \emptyset \Longrightarrow_{KB} E_1, R_1 \Longrightarrow_{KB} \cdots$$

is fair, then whenever $u \xleftrightarrow{\;*\;}_{E_i \cup R_i} v$ for some i, we have a rewrite proof

$$u \xrightarrow{\;*\;}_{R^\infty} w \xleftarrow{\;*\;}_{R^\infty} v.$$

This theorem shows that a completion procedure can be used as a semi-decision procedure for the word problem for an arbitrary E. Thus, a fair completion procedure will either fail, terminate with success, or run forever in creating an *infinite* canonical set of rewrite rules equivalent to E. The procedure can halt with success if at some stage n, $E_n = \emptyset$, R^n is reduced, and each of the critical pairs in $cp(R^n)$ has already been generated and appeared in some E_i for $i < n$; in this case, R_n is a completion of E. It turns out that fair completion sequences are not too hard to generate, and so the crucial problem turns out to be finding an ordering which can orient all the equations and critical pairs generated. In most current implementations, this involves human intervention to orient rules which can not otherwise be oriented by the given term ordering. The interested reader is referred to the works mentioned above for further details.

An important consequence of Theorem 3.6.4 is that by using the notion of *ground confluence* and the fact that reduction orderings total on ground terms always exist, it is possible to avoid failure due to equations which can not be oriented. This extension of completion, called *unfailing completion*, or *ordered completion*, was developed by Bachmair, Dershowitz, Hsiang, and Plaisted [8, 12, 9], and turns out to be crucial in the completeness results of Chapter §6; we shall present the necessary results in a simplified form in Section §6.1.

CHAPTER 4

E-UNIFICATION

In Section §3.3 we defined the standard unification of terms, most general unifiers, and showed how the abstract non-deterministic method of transformations on systems of equations provides a procedure for unification which either fails or terminates with an explicit representation of the *mgu* of the original system. The notion of standard unification is based on making two (first-order) terms syntactically identical, but in fact, we could generalize this to any relation P on terms, defining "P-unification" to be the problem of determining for two terms u and v if there exists some substitution θ such that $(\theta(u), \theta(v)) \in P$. In this section we present the basic notions of E-unification, where this relation P is represented by a finite set of equations E. The two following chapters will present a general procedure for E-unification via the method of transformations; later in this monograph, in Chapter §7, we present a generalization of unification to higher-order terms, and develop a non-deterministic procedure in the same fashion.

4.1 Basic Definitions and Results

Definition 4.1.1 Let E be a finite set of equations. We say that a substitution θ is a unifier of an equation $s \approx t$ *modulo* E, or an E-*unifier of s and t*, iff $\theta(s) \overset{*}{\longleftrightarrow}_E \theta(t)$. A substitution θ is an E-unifier of a system S if it E-unifies every equation in S, and the set of all such E-unifiers will be denoted $U_E(S)$. If $S = \{s \approx t\}$, then this will be denoted by $U_E(s, t)$.

Unfortunately, unification modulo a set of equations does not enjoy the nice properties of standard unification. Since the word problem for an arbitrary E is an instance of the E-unification problem, E-unification is undecidable, but since we can dovetail the enumeration of all possible substitutions and equational proofs, the set of E-unifiers of any two terms is always recursively enumerable. Another difference from standard unification is that most general unifiers do not necessarily exist. In fact, it is

possible that two terms have an infinite set of independent E-unifiers, as
we now show.

Example 4.1.2 Let $\Sigma = \{\cdot\} \cup \Sigma_0$, where Σ_0 contains at least one
constant symbol "a", and, using infix notation, let $E = \{(x' \cdot y') \cdot z' \doteq x' \cdot (y' \cdot z')\}$. This set axiomatizes non-empty strings over the set of constant
symbols Σ_0, and so we represent terms as simply strings of constants and
variables. Now consider the problem of E-unifying the two "strings" ax
and xa. If $X_1 \ldots X_n$ is an E-unifier, for $X_i \in \Sigma_0 \cup X$ and $n \geq 1$, then since
$aX_1 X_2 \ldots X_n = X_1 X_2 \ldots X_n a$ we must have $a = X_1 = X_2 = \ldots X_n = a$,
and so

$$U_E(ax, xa) = \{[a/x], [aa/x], [aaa/x], \ldots\}.$$

Clearly this set is infinite, and none of the substitutions subsumes any
other, since they are all ground, and so no most general E-unifier, nor even
a finite set of "more general E-unifiers," can exist. We now discuss some
notions needed to deal with this more complex situation.

Definition 4.1.3 Given a finite set E of equations and any set V of
variables, we say that two substitutions σ and θ are *equal modulo E over*
V, denoted by $\sigma =_E \theta[V]$, iff $\forall x \in V$, $\sigma(x) \stackrel{*}{\longleftrightarrow}_E \theta(x)$. We say that σ is
more general modulo E than θ over V, denoted by $\sigma \leq_E \theta[V]$, iff there
exists some substitution η such that $\theta =_E \sigma \circ \eta[V]$. When V is the set of
all variables, we drop the notation $[V]$, and similarly we drop the subscript
E when $E = \emptyset$.

An important property of the relation $=_E$ which will be needed later is
given by

Lemma 4.1.4 If $\theta =_E \sigma$ then for any system S, $\theta \in U_E(S)$ iff $\sigma \in U_E(S)$.

Proof. For any equation $u \approx v$ in S, a simple induction on the structure
of u and v suffices to show that $\theta(u) \stackrel{*}{\longleftrightarrow}_E \theta(v)$ iff $\sigma(u) \stackrel{*}{\longleftrightarrow}_E \sigma(v)$. $\square$

From this lemma and the stability of E-congruence we can show

Corollary 4.1.5 If $\sigma \in U_E(S)$ and $\sigma \leq_E \theta[Var(S)]$ then $\theta \in U_E(S)$.

Note that this result is true in the special case that $\sigma \leq \theta[Var(S)]$.
Now we generalize the concept of a $mgu(S)[V]$ to sets of E-unifiers; this

formulation of a generating set for a set of E-unifiers is due to [131]; we present a modification of the definition from [43] for equation systems.[1]

Definition 4.1.6 Given a finite set E of equations, a system S, and a finite set V of 'protected' variables, a set U of substitutions is a *complete set of E-unifiers for S away from V* (which we shall abbreviate by $CSU_E(S)[V]$) iff
  (i) For all $\sigma \in U$, $D(\sigma) \subseteq Var(S)$ and $I(\sigma) \cap (V \cup D(\sigma)) = \emptyset$;
  (ii) $U \subseteq U_E(S)$;
  (iii) For every $\theta \in U_E(S)$, there exists some $\sigma \in U$ such that $\sigma \leq_E \theta[Var(S)]$.
The first condition is called the *purity condition*, the second the *coherence condition*, and the last the *completeness condition*. If S consists of a single equation $u \approx v$ then we use the abbreviation $CSU_E(u, v)[V]$. When the use of V is not relevant to our discussion we shall drop the notation $[V]$.

We now justify the purity condition and show the generality of idempotent E-unifiers.

Lemma 4.1.7 For any system S, substitution θ, and set of protected variables W, if $\theta \in U_E(S)$ then there exists some substitution σ such that
  (i) $D(\sigma) \subseteq Var(S)$ and $I(\sigma) \cap (W \cup D(\sigma)) = \emptyset$;
  (ii) $\sigma \in U_E(S)$;
  (iii) $\sigma \leq \theta[Var(S)]$ and $\theta \leq \sigma[Var(S)]$.

Proof. If $\sigma = \theta|_{Var(S)}$ satisfies condition (i), then we have our result trivially. Otherwise, if $I(\theta) = \{x_1, \ldots, x_n\}$ then let $\{y_1, \ldots, y_n\}$ be a set of new variables disjoint from the variables in W, $D(\theta)$, $I(\theta)$, and $Var(S)$. Now define the renaming substitutions $\rho_1 = [y_1/x_1, \ldots, y_n/x_n]$ and $\rho_2 = [x_1/y_1, \ldots, x_n/y_n]$, and then let $\sigma = \theta \circ \rho_1|_{Var(S)}$. Clearly σ satisfies (i), and since $\sigma = \theta \circ \rho_1[Var(S)]$, we have the second part of (iii). Now since $\rho_1 \circ \rho_2 = Id[Var(S) \cup I(\theta)]$, we must have $\theta = \theta \circ \rho_1 \circ \rho_2[Var(S)]$. But then by the fact that $\sigma = \theta \circ \rho_1[Var(S)]$ we have $\theta = \sigma \circ \rho_2[Var(S)]$, proving the first part of (iii). To show (ii), observe that for any $u \approx v \in S$ we have $\theta(u) \overset{*}{\longleftrightarrow}_E \theta(v)$, and so by the stability of E-congruence we have

$$\sigma(u) = \rho_1(\theta(u)) \overset{*}{\longleftrightarrow}_E \rho_1(\theta(v)) = \sigma(v),$$

which shows that $\sigma \in U_E(S)$. $\square$

---

[1] We also generalize slightly the Fages and Huet definition by allowing the protected set of variables to be arbitrary. The original definition imposed the restriction that $V \cap Var(S) = \emptyset$ in order that variable renaming not be necessary. We relax this restriction so that we have a true generalization of a $mgu(S)[V]$ to E-unifiers, and allow renaming to be imposed or not, by setting V appropriately.

This shows us that for any S and W, the set of all unifiers satisfying condition (i) and (ii) of Definition 4.1.6 is a $CSU(S)[W]$, and so in particular there is no loss of generality in considering only idempotent E-unifiers in what follows. This will simplify several of the definitions and proofs.

It would be desirable, as in the case of $E = \emptyset$, to show the existence of complete sets of E-unifiers satisfying some minimality conditions. Such conditions were first proposed by Huet in the framework of higher-order unification [72].

Definition 4.1.8   Two minimality conditions may be defined. Let s and t be two terms, and U a complete set of unifiers for s and t.
  (i) (Minimality) For any two substitutions σ, $\theta \in U$, if $\sigma \leq_E \theta[V]$ then $\sigma = \theta$;
 (ii) (Non-congruence) For any two substitutions σ, $\theta \in U$, if $\sigma \equiv_E \theta[V]$ then $\sigma = \theta$.

Note that minimality implies non-congruence. Unfortunately, there are difficulties with these concepts. Both are non-recursively enumerable, and minimality cannot always be achieved: there exists a set of equations E and two terms s and t such that there is *no* complete and minimal set of E-unifiers for s and t.[2] Thus, the notions of completeness and minimality may conflict.

4.2 Methods for E-Unification

To date there have been two basic approaches to the problem of generating complete sets of E-unifiers, neither one valid for arbitrary sets E. The first is to examine a particular equational theory, such as the theory of semigroups, and develop an ad-hoc method for the problem. This was the approach taken in the pioneering work of [131], and extended to a large number of specific theories since then. Many of these specialized theories have very interesting properties, e.g., the E-unification problem may be decidable, or it may be possible always to find finite CSUs. A good survey of results of this sort may be found in [145] or [83]. In this book we are interested in more general forms of E-unification, and so we shall not discuss this ad-hoc approach further.

The second approach developed for E-unification is called *narrowing* and depends on the special characteristics of canonical sets of rewrite rules.

---

[2] This phenomenon was first noticed in higher-order unification (see [72]), and extended to E-unification in [43]; for further results, see [7] and the references presented there.

Narrowing was first presented in [146] and [100], but the E-unification algorithm based on this technique first appeared in [46] and was refined by [78]. Since then the basic method has been developed by various researchers [79,85,91,48,111,120,136]. A survey of some of the important results concerning the method can be found in [93].

Since this approach applies to an important class of theories and forms the background to the more general results presented in the next two chapters, we shall present the technical results of the method in detail. First we define the notion of a normalized substitution, and then define the *one step narrowing relation*. For all the results and definitions in the rest of this section we assume that R is a canonical set of rewrite rules. We shall therefore use the phrase 'R-unifier' in place of the previous 'E-unifier.'

Definition 4.2.1 A substitution θ is called *normalized with respect to R* or *reduced w.r.t. R* if for every x in the support of θ, $\theta(x)$ is in normal form with respect to R. (If R is available from context, we simply say θ is *normalized* or *reduced*.)

A consequence of this is that if $l \rightarrow r \in R$ and $\theta(t) \longrightarrow_{[\alpha, l \rightarrow r, \rho]} t'$, where θ is normalized, then $\alpha \in NonVarDom(t)$.

Definition 4.2.2 The *one step narrowing relation* $\succ\!\!\longrightarrow_R$ is defined so that for any terms s and t and set W of protected variables, we have

$$s \succ\!\!\longrightarrow_{[\alpha, l \rightarrow r, \sigma, W]} t$$

iff

(i) $\alpha \in NonVarDom(s)$,
(ii) $\sigma = mgu(s/\alpha, l)$, and
(iii) $t = \sigma(s[\alpha \leftarrow r])$,

for some variant $l \rightarrow r$ of a rewrite rule in R. In this case, we say that t is a *one-step narrowing of s*. (Narrowing is sometimes also called *surreduction*.)

Without loss of generality, we assume in what follows that the *mgus* in narrowing sequences are such as would be produced by the set $\mathcal{ST}$. Before we present the fundamental result which shows the relationship between normalized substitutions, rewriting, and one-step narrowing, we give a lemma which is useful for the main proof.

Lemma 4.2.3 If s and t are two terms such that $Var(s) \cap Var(t) = \emptyset$ and if σ is a *mgu* of s and t such that $D(\sigma) \subseteq Var(s) \cup Var(t)$, then $I(\sigma) = Var(\sigma(s)) = Var(\sigma(t))$.

Proof. Since $\sigma(s) = \sigma(t)$ we must have $Var(\sigma(s)) = Var(\sigma(t))$, and since $D(\sigma) \subseteq Var(s) \cup Var(t)$ we have $I(\sigma) \subseteq Var(\sigma(s)) = Var(\sigma(t))$. Now suppose there is some $x \in Var(\sigma(s)) = Var(\sigma(t))$ such that $x \notin I(\sigma)$. Then $x \in Var(s) \cap Var(t)$, contradicting our assumption. Therefore $Var(\sigma(s)) = Var(\sigma(t)) \subseteq I(\sigma)$. $\quad\square$

The central result which shows the precise manner in which we may 'lift' rewrite steps to narrowing steps was first shown by [78], and was presented in detail in [93]. We present a modified version of this result.

Lemma 4.2.4 (Lifting Lemma for Narrowing) Let R be a canonical set of rewrite rules, u be a term, W be a set of 'protected variables' containing $Var(u)$, θ be a normalized substitution such that $D(\theta) \subseteq W$, and suppose $\theta(u) \longrightarrow_{[\alpha,l \dot{\to} r,\rho]} v$ is a reduction, where $l \dot{\to} r$ is a variant of a rewrite rule in R (i.e., the variables in $Var(l)$ are *new* and occur *only* in this rule, and $Var(l) \cap W = \emptyset$) and, w.l.g., $D(\rho) = Var(l)$. Then there exists a set W' extending W, a narrowing step $u \succ\!\!\longrightarrow_{[\alpha,l \dot{\to} r,\sigma,W]} v'$, and a substitution θ' such that

(i) $Var(v') \subseteq W'$ and $D(\theta') \subseteq W'$,
(ii) $\theta = \sigma \circ \theta'[W]$,
(iii) $v = \theta'(v')$, and
(iv) θ' is normalized.

which can be illustrated by the following figure:

$$
\begin{array}{ccc}
\theta(u) & \xrightarrow{\quad[\alpha,l \dot{\to} r,\rho]\quad} & v \\[2pt]
\Big\Uparrow \theta & & \Big\Uparrow \theta' \\[2pt]
u & \succ\!\!\!\longrightarrow_{[\alpha,l \dot{\to} r,\sigma,W]} & v'
\end{array}
$$

Proof. Since θ is normalized, no rewrite can occur in any term $\theta(x)$ for $x \in D(\theta)$, so that $\theta(u)/\alpha = \theta(u/\alpha)$. But since $l \dot{\to} r$ is a variant,

$$\theta \cup \rho(u/\alpha) = \theta(u/\alpha) = \rho(l) = \theta \cup \rho(l),$$

and thus there exists a $\sigma = mgu(u/\alpha, l)[W \cup Var(l)]$ such that $D(\sigma) = Var(u/\alpha) \cup Var(l)$ and there exists some η such that (by Corollary 3.3.12) $\theta \cup \rho = \sigma \circ \eta[W \cup Var(l)]$ and since σ is idempotent by definition, we can assume that $D(\sigma) \cap D(\eta) = \emptyset$. (Thus, $D(\eta) \cap Var(u/\alpha, l) = D(\eta) \cap D(\rho) = \emptyset$.) Since σ is also an $mgu(u/\alpha, l)[W]$ we can define the narrowing step $u \succ\!\!\longrightarrow_{[\alpha,l \dot{\to} r,\sigma,W]} v'$ where $v' = \sigma(u[\alpha \leftarrow r])$ and $Var(v') \subseteq (Var(u) - D(\sigma)) \cup I(\sigma)$. We define the new set of protected variables

$$W' = W \cup I(\sigma).$$

Since $Var(u) \subseteq W$ then we must have $Var(v') \subseteq W'$, proving the first part of (i).

Now let $\theta' = \eta|_{W'}$, so that the second part of (i) is immediate. Now, since $D(\rho)$ is disjoint from W and $I(\theta)$, and by the definition of θ' and of W', we have

$$\theta = \theta \cup \rho = \sigma \circ \eta = \sigma \circ \theta'[W],$$

proving (ii).

To show (iii) we first observe that by (ii) we must have $\theta(u) = \theta'(\sigma(u))$. Also, by the definition of θ', the fact the $l \to r$ is a variant, and because $D(\eta) \cap Var(l) = \emptyset$, we must have $\rho = \sigma \circ \eta = \sigma \circ \theta'[Var(l)]$, and since $Var(r) \subseteq Var(l)$ then $\rho(r) = \theta'(\sigma(r))$. Therefore

$$\begin{aligned}
v &= \theta(u)[\alpha \leftarrow \rho(r)] \\
&= \theta'(\sigma(u))[\alpha \leftarrow \theta'(\sigma(r))] \\
&= \theta'(\sigma(u[\alpha \leftarrow r])) \\
&= \theta'(v').
\end{aligned}$$

Finally, to prove (iv) we must show that for any x in $D(\theta')$, $\theta'(x)$ is irreducible. Now the fact that $D(\theta') \subseteq W' = W \cup I(\sigma)$ gives us two cases for any $x \in D(\theta')$. If $x \in W$, then, since $D(\theta') \cap D(\sigma) = \emptyset$ and $\theta = \sigma \circ \theta'[W]$ we have $\theta'(x) = \theta'(\sigma(x)) = \theta(x)$ and so $\theta'(x)$ must be irreducible, because θ is normalized. Otherwise, if $x \in I(\sigma)$, then by the previous lemma we have $I(\sigma) = Var(\sigma(u/\alpha))$ and hence, for every $x \in I(\sigma) \cap D(\theta')$, there exists a $z \in Var(u/\alpha) \subseteq W$ such that $\theta(z) = \theta'(\sigma(z))$ and $\theta'(x) = \theta(z)/\beta$, where β is the address of x in $\sigma(z)$. Since $\theta(z)$ is irreducible, so are all its subterms, including $\theta'(x)$, and we are done. $\qquad\square$

The use of the sets of protected variables W and W' is a technical necessity in order to use this lemma in the induction step of the completeness proof below. The fact that $\theta = \sigma \circ \theta'[W]$ in this lemma shows us that the narrowing substitution σ contains a 'piece' of the substitution θ, i.e. if we let $\sigma' = \sigma|_{Var(u)}$ then $\sigma' \leq \theta$. The narrowing procedure for R-unification iterates this process in order to incrementally build up an R-unifier.

Recall that if R is canonical, then there is a simple decision procedure for the word problem for R; to test if $s \xleftrightarrow{*}_R t$ we check if $s{\downarrow} = t{\downarrow}$, i.e., by finding a rewrite proof $s \xrightarrow{*}_R w \xleftarrow{*}_R t$ for some w. Now if we let eq be a *new* function symbol not occurring in Σ, this can be collapsed into a more convenient form using a single rewrite sequence

$$eq(s,t) \longrightarrow_R eq(s_1,t_1) \longrightarrow_R \ldots \longrightarrow_R eq(s_n,t_n),$$

where $s_n = t_n$ and no rewrite steps can take place at the root. Therefore, if θ is an R-unifier of two terms s and t, i.e., $\theta(s) \xleftrightarrow{*}_R \theta(t)$, then there must exist some rewrite sequence

$$\theta(eq(s,t)) = eq(s_0', t_0') \longrightarrow_{[l_1 \to r_1]} eq(s_1', t_1') \ldots \longrightarrow_{[l_n \to r_n]} eq(s_n', t_n')$$

for some $n \geq 0$, where $s_n' = t_n'$.

The method of narrowing basically searches the *narrowing tree* of all narrowing sequences originating from two terms to be R-unified, in order to find a narrowing sequence which simulates a rewrite proof (as above) which proves that the two terms are R-unifiable. Before we present the technical results which show the soundness and completeness of this method, we give an illustration.

Example 4.2.5 Let $\Sigma = \{a, \cdot\}$ (where "$\cdot$" represents concatenation of strings), $R = \{x_1(y_1 z_1) \to (x_1 y_1)z_1\}$, and consider the terms ax and xa. Since $[a/x]$ is a *mgu* of ax and xa, then $[a/x]$ is an R-unifier of ax and xa (this was found after a narrowing sequence of length 0). Now consider the narrowing sequence

$$eq(ax, xa) \rightarrowtail_{[1, x_1(y_1 z_1) \to (x_1 y_1)z_1, \sigma]} eq((ay_1)z_1, (y_1 z_1)a),$$

where $\sigma = [a/x_1, y_1 z_1/x]$ and the final term has a *mgu* $\mu = [a/y_1, a/z_1]$. Then $\theta = \sigma \circ \mu = [a/x_1, aa/x, a/y_1, a/z_1]$ is an R-unifier of ax and xa, as shown by the following rewrite sequence:

$$\theta(eq(ax, xa)) = eq(a(aa), (aa)a) \longrightarrow_{[1, x_1(y_1 z_1) \to (x_1 y_1)z_1, \rho]} eq((aa)a, (aa)a),$$

where $\rho = [a/x_1, a/y_1, a/z_1]$.

We now show that the result of narrowing two terms until a (standard) unifiable equation is obtained, and then concatenating the *mgu* found with all the narrowing substitutions generated, always results in an R-unifier of the original two terms.

Theorem 4.2.6 (Soundness of Narrowing) Let R be a canonical set of rewrite rules, s and t be two terms in $T_\Sigma(X)$ and eq be a new function symbol not in Σ. For any narrowing sequence

$$eq(s,t) \rightarrowtail_{[l_1 \to r_1, \sigma_1]} eq(s_1, t_1) \rightarrowtail_{[l_2 \to r_2, \sigma_2]} \cdots \rightarrowtail_{[l_n \to r_n, \sigma_n]} eq(s_n, t_n),$$

such that each $l_i \to r_i$ is a variant of a rule in R and s_n and t_n are (standard) unifiable, the substitution

$$\theta = \sigma_1 \circ \ldots \circ \sigma_n \circ \mu,$$

is an R-unifier of s and t, where $\mu = mgu(s_n, t_n)$.

Proof. (By induction on n) If $n = 0$ then the result is trivial. If $n > 0$ then by the induction hypothesis, we have that $\theta' = \sigma_2 \circ \ldots \circ \sigma_n \circ \mu$ is an R-unifier of s_1 and t_1, i.e., $\theta'(s_1) \xleftrightarrow{*}_R \theta'(t_1)$. Now, w.l.g., assume that the first narrowing step takes place inside u (note that it can not take place at the root, since $eq \notin \Sigma$), i.e. $s \rightarrowtail_{[l_1 \dot{\to} r_1, \sigma_1]} s_1$ and $t_1 = \sigma_1(t)$, so then by the definition of one-step narrowing, we have $\sigma_1(s) \longrightarrow_{[l_1 \dot{\to} r_1, \sigma_1]} s_1$, with the result that by the stability of $\longrightarrow_R$ we have

$$\theta(s) = \theta'(\sigma_1(s)) \longrightarrow_R \theta'(s_1) \xleftrightarrow{*}_R \theta'(t_1) = \theta'(\sigma_1(t)) = \theta(t).$$
$\square$

The next result uses the lifting lemma to show that any time a substitution θ R-unifies two terms, it is always possible to find a narrowing sequence which generates a substitution more general than θ.

Theorem 4.2.7 (Completeness of Narrowing) Let R be a canonical set of rewrite rules, s and t be two terms in $T_\Sigma(X)$ and eq a new function symbol not in Σ. For any substitution θ, if $\theta(s) \xleftrightarrow{*}_R \theta(t)$ then for any set of protected variables W containing $Var(u,v)$ and $D(\theta)$ there exists a narrowing sequence

$$eq(s,t) \rightarrowtail_{[l_1 \dot{\to} r_1, \sigma_1]} eq(s_1, t_1) \rightarrowtail_{[l_2 \dot{\to} r_2, \sigma_2]} \cdots \rightarrowtail_{[l_n \dot{\to} r_n, \sigma_n]} eq(s_n, t_n),$$

for some $n \geq 0$, and a *mgu* μ of s_n and t_n such that

$$\sigma_1 \circ \ldots \circ \sigma_n \circ \mu \leq_R \theta[W].$$

Proof. First of all, let θ' be the normalized version of θ, i.e., for every $x \in D(\theta)$, let $\theta'(x) = \theta(x)\!\downarrow$. Thus, for every $x \in D(\theta)$ we have $\theta(x) \xrightarrow{*}_R \theta'(x)$ and so $\theta =_R \theta'$ and $\theta(u) \xrightarrow{*}_R \theta'(u)$ for any term u. Therefore

$$\theta'(s) \xleftarrow{*}_R \theta(s) \xleftrightarrow{*}_R \theta(t) \xrightarrow{*}_R \theta'(t)$$

and thus

$$\theta'(s) \xrightarrow{*}_R w \xleftarrow{*}_R \theta'(t)$$

for some w, so that we must have some rewrite sequence

$$\theta'(eq(s,t)) = eq(s'_0, t'_0) \longrightarrow_{[l_1 \dot{\to} r_1]} eq(s'_1, t'_1) \ldots \longrightarrow_{[l_n \dot{\to} r_n]} eq(s'_n, t'_n)$$

where $s'_n = t'_n$. We proceed by induction on n to show that for any such rewrite sequence where θ' is normalized and for any set W containing $Var(s,t)$ and $D(\theta')$, there exists the corresponding narrowing sequence

$$eq(s,t) \rightarrowtail_{[l_1 \dot{\to} r_1, \sigma_1]} eq(s_1, t_1) \rightarrowtail_{[l_2 \dot{\to} r_2, \sigma_2]} \cdots \rightarrowtail_{[l_n \dot{\to} r_n, \sigma_n]} eq(s_n, t_n),$$

for some $n \geq 0$, and a *mgu* μ of s_n and t_n such that $\sigma_1 \circ \ldots \circ \sigma_n \circ \mu \leq \theta'[W]$, from which our result will follow. If $n = 0$ then $s_0' = t_0'$ and so s and t are unifiable, and thus if $\mu = mgu(s,t)[W]$ then by Corollary 3.3.12 we have $\mu \leq \theta'[W]$.

Otherwise, if $n > 0$ then by the lifting lemma, there is some narrowing step

$$eq(s,t) \rightarrowtail_{[l_1 \dot{\rightarrow} r_1, \sigma_1, W]} eq(s_1, t_1)$$

and new set of protected variables $W' = W \cup I(\sigma_1)$ and some normalized substitution θ'' such that $D(\theta''), Var(eq(s_1, t_1)) \subseteq W'$, $\theta' = \sigma_1 \circ \theta''[W]$, and $eq(s_1', t_1') = \theta''(eq(s_1, t_1))$. But then we may apply the induction hypothesis to the rewrite sequence

$$\theta''(eq(s_1, t_1)) = eq(s_1', t_1') \longrightarrow_{[l_2 \dot{\rightarrow} r_2]} \cdots \longrightarrow_{[l_n \dot{\rightarrow} r_n]} eq(s_n', t_n'),$$

and the set W' to obtain the corresponding narrowing sequence

$$eq(s_1, t_1) \rightarrowtail_{[l_2 \dot{\rightarrow} r_2, \sigma_2]} \cdots \rightarrowtail_{[l_n \dot{\rightarrow} r_n, \sigma_n]} eq(s_n, t_n),$$

and a *mgu* μ of s_n and t_n such that $\sigma_2 \circ \ldots \circ \sigma_n \circ \mu \leq \theta''[W']$. By adding the narrowing step $eq(s,t) \rightarrowtail_{[l_1 \dot{\rightarrow} r_1, \sigma_1]} eq(s_1, t_1)$ to the front of this sequence, we have our result, since by our choice of W' we must have

$$\sigma_1 \circ \sigma_2 \circ \ldots \circ \sigma_n \circ \mu \leq \sigma_1 \circ \theta'' = \theta'[W].$$

Finally, since $\theta =_R \theta'$, we must have

$$\sigma_1 \circ \sigma_2 \circ \ldots \circ \sigma_n \circ \mu \leq_R \theta[W].$$

$\square$

For any terms u and v and set of variables W let us call a sequence

$$eq(u,v) \rightarrowtail_{[l_1 \dot{\rightarrow} r_1, \sigma_1]} eq(u_1, v_1) \rightarrowtail_{[l_2 \dot{\rightarrow} r_2, \sigma_2]} \cdots \rightarrowtail_{[l_n \dot{\rightarrow} r_n, \sigma_n]} eq(u_n, v_n)$$

associated with a set of variables W, as shown in the previous result, a *narrowing sequence away from* W and denote this by

$$eq(u,v) \overset{*}{\rightarrowtail}_{[\theta, W]} eq(u_n, v_n),$$

where $\theta = \sigma_1 \circ \ldots \circ \sigma_n \circ \mu$ for some mgu μ of u_n and v_n, as above. We may sum up the results of this section with the following theorem.

Theorem 4.2.8 For any canonical set of rewrite rules R, any terms s and t, and set of protected variables W containing $Var(s,t)$, the set

$$\{\, \theta|_{Var(s,t)} \mid eq(s,t) \overset{*}{\rightarrowtail}_{[\theta,W]} eq(s',t') \,\}$$

is a $CSU_R(s,t)[W]$.

Proof. The soundness and completeness criteria have just been shown. For the purity condition, we need only show that each such θ is away from W, and so we proceed by induction on the length of a given sequence

$$eq(u,v) \rightarrowtail_{[l_1 \to r_1,\sigma_1]} eq(u_1,v_1) \rightarrowtail_{[l_2 \to r_2,\sigma_2]} \cdots \rightarrowtail_{[l_n \to r_n,\sigma_n]} eq(u_n,v_n)$$

to show that $(I(\sigma_1) \cup \ldots \cup I(\sigma_n) \cup I(\mu)) \cap W = \emptyset$, for any given W, from which the result will hold. For $n = 0$ the result is trivial, since μ will be a $mgu(u,v)[W]$. Otherwise, for $n > 0$, we have $I(\sigma_1) \cap W = \emptyset$ by definition, and then we choose $W' = W \cup I(\sigma)$ and apply the induction hypothesis to get $(I(\sigma_2) \cup \ldots \cup I(\sigma_n) \cup I(\mu))$ disjoint from W', and since $W \subseteq W'$, we are done. $\square$

This result gives us a complete strategy for E-unification in the case that E is equivalent to a canonical set of rewrite rules R; we simply search the *narrowing tree* of all possible narrowing sequences from the term $eq(s,t)$ in some complete fashion (say breadth-first) and whenever a unifiable equation is found, return the composition of all the narrowing substitutions on the path back to the root term.

The fundamental idea behind the narrowing procedure is the *lifting* of rewrite proofs to narrowing sequences. The interesting feature of canonical rewrite systems is of course that rewriting is non-deterministic, so that *any* strategy (e.g. top-down, bottom-up, inner-most left-most) for rewriting will reduce a term to its normal form. It turns out that by examining the result of lifting rewrite proofs found under various strategies, we can improve the narrowing procedure by reducing the search space without sacrificing completeness. There are two principal improved versions of narrowing which have been defined. The first, *basic narrowing*, due to [78], reduces the search space by forbidding narrowing at addresses in that part of the term introduced by the narrowing substitution, and is a lifting of an *innermost* rewrite proof (see Definition 6.1.11 and Lemma 6.1.12). The second, *normalized narrowing*, due to [46], reduces the search space by normalizing a term before applying a narrowing step, and is a lifting of (roughly) a certain kind of *top-down* rewrite proof. Both these restrictions reduce the size of the narrowing tree and are complete, but unfortunately, as discussed in [136], the naive combination of these two is not complete. Since these results will not be used in this monograph, we shall not discuss them further, but instead refer the interested reader to the references just mentioned.

The major problem with this technique for E-unification is that it is of course only usable when E is in fact a canonical set of rewrite rules or when the theory can be compiled into a set of canonical rewrite rules as discussed in the previous section on completion. This process is undecidable in general, and, even if the theory is completable, may require human intervention to determine a termination ordering, perhaps by explicitly ordering rewrite rules during the completion process. Thus, not only are there *two* levels of undecidability in such a system, first attempting to complete the set E to a canonical set of rewrite rules, and then performing narrowing, but the current approaches to the completion phase often need human intervention. The extensions to narrowing which account for some of the cases where the completion procedure fails, such as when E contains a commutative axiom, begin to look rather ad-hoc again, and so this approach seems rather unsuitable as a basis for a theory of E-unification and as a general paradigm in the larger context of automated reasoning. In the next two chapters we examine the more general problem of E-unification in arbitrary theories, and show how our abstract approach subsumes the technique of narrowing.

CHAPTER 5

E-Unification via Transformations

We now show how to extend the set of transformations $\mathcal{ST}$ given in Section §3.3 to perform E-unification of a system under some arbitrary E, and develop the non-deterministic completeness of the method using a new formalism for 'proofs' that two terms are E-unifiable, known as *equational proof trees*. The new set of transformations is fully general in that it is capable of enumerating a $CSU_E(S)$ for any system S and set of equations E, and we intend this chapter to provide a paradigm for the abstract study of complete methods for general E-unification. The set of E-unifiers found by this method is highly redundant, however, and in the next chapter, we show how to restrict this method to avoid rewriting at variable occurrences while still retaining the ability to enumerate a $CSU_E(S)$.

5.1 The Set of Transformations $\mathcal{BT}$

We shall follow for the most part the plan of Section §3.3, in order to highlight the essential similarities and differences between standard unification and E-unification. First we examine the significance of solved form systems in this new context.

Lemma 5.1.1 If $S' = \{x_1 \approx t_1, \ldots, x_n \approx t_n\}$ is a system in solved form, then $\{\sigma_{S'}\}$ is a $CSU_E(S')[V]$ for any V such that $V \cap Var(S') = \emptyset$.

Proof. The first two conditions in Definition 4.1.6 are satisfied, since $\sigma_{S'}$ is an idempotent mgu of S', $V \cap Var(S') = \emptyset$, and $I(\sigma_{S'}) \subseteq Var(S')$. Now, if $\theta \in U_E(S')$, then $\theta =_E \sigma_{S'} \circ \theta$, since $\theta(x_i) \overset{*}{\longleftrightarrow}_E \theta(t_i) = \theta(\sigma_{S'}(x_i))$ for $1 \leq i \leq n$, and $\theta(x) = \theta(\sigma_{S'}(x))$ otherwise. Thus $\sigma_{S'} \leq_E \theta$ and so obviously $\sigma_{S'} \leq_E \theta[Var(S')]$. $\square$

This allows us to effectively ignore any E-unifiers which use rewrite steps between equations in solved systems, if we are just interested in complete sets of unifiers.

We may analyse the process of finding a $CSU_E(u, v)$ for two terms u

and v as follows. If $\theta \in U_E(u, v)$ then there must exist some sequence

$$\theta(u) = u_0 \longleftrightarrow_{[\alpha_1, l_1 \doteq r_1, \rho_1]} u_1$$
$$\longleftrightarrow_{[\alpha_2, l_2 \doteq r_2, \rho_2]} u_2 \cdots$$
$$\longleftrightarrow_{[\alpha_m, l_m \doteq r_m, \rho_m]} u_m = \theta(v)$$

with m minimal (so that there are no redundant steps), $D(\rho_i) \subseteq Var(l_i, r_i)$ for $1 \leq i \leq m$. Since all the equations are variants, then we can assume that $D(\theta), D(\rho_1), \ldots, D(\rho_m)$ are pairwise disjoint, and we can form an *extended E-unifier* $\theta' = \theta \cup \rho_1 \cup \ldots \cup \rho_m$, so that we have

$$\theta'(u) = u_0 \longleftrightarrow_{[\alpha_1, l_1 \doteq r_1, \theta']} u_1$$
$$\longleftrightarrow_{[\alpha_2, l_2 \doteq r_2, \theta']} u_2 \cdots$$
$$\longleftrightarrow_{[\alpha_m, l_m \doteq r_m, \theta']} u_m = \theta'(v).$$

Given any such rewrite sequence and extended E-unifier, we have several cases.

(1) $m = 0$ and $\theta' = \theta \in U(u, v)$. Then the analysis for standard unification is sufficient.

(2) $m \neq 0$ and some rewrite step occurs at the root of some u_i. Assume that if one of u, v is not a variable, it is u, and pick the left-most rewrite step; then

$$\theta'(u) \xleftrightarrow{*}_E \theta'(l_i) \longleftrightarrow_{[\epsilon, l_i \doteq r_i, \theta']} \theta'(r_i) \xleftrightarrow{*}_E \theta'(v),$$

for some i, $1 \leq i \leq m$, where there is no rewrite at the root between $\theta'(u)$ and $\theta'(l_i)$.

(3) $m \neq 0$ and no rewrite step occurs at the root of any u_i.

(a) $u = f(u_1, \ldots, u_n)$, $v = f(v_1, \ldots, v_n)$ for some $f \in \Sigma_n$ with $n > 0$, and therefore $\theta'(u_i) \xleftrightarrow{*}_E \theta'(v_i)$ for $1 \leq i \leq n$.

(b) Either u or v is a variable; assume u is a variable.

    (i) $v = f(v_1, \ldots, v_n)$ for some $f \in \Sigma_n$ with $n > 0$, $\theta'(u) = f(t_1, \ldots, t_n)$ for some terms $t_1, \ldots, t_n$, and thus $t_i \xleftrightarrow{*}_E \theta'(v_i)$ for $1 \leq i \leq n$.

    (ii) v is a variable and $\theta'(u) = f(t_1, \ldots, t_n)$ and $\theta'(v) = f(t'_1, \ldots, t'_n)$ for some terms $t_1, \ldots, t_n, t'_1, \ldots, t'_n$, where $t_i \xleftrightarrow{*}_E t'_i$ for $1 \leq i \leq n$.

By recursively applying this analysis to the subsequences found in each case, every rewrite step in the original sequence can be accounted for. We use cases (2) and (3) to define two new transformation rules to account for the presence of rewrite steps in a unification problem.

Definition 5.1.2 (The set of transformation rules $\mathcal{BT}$) To the transformations $\mathcal{ST}$ we add two more to deal with equations.

Root Rewriting: Let $u \approx v$ be an equation and if one of u or v is not a variable, assume that it is u. Then

$$\{u \approx v\} \cup S \Longrightarrow \{u \approx l, r \approx v\} \cup S, \tag{4}$$

where $l \doteq r$ is an alphabetic variant of an equation in $E \cup E^{-1}$ such that $Var(l, r) \cap (Var(S) \cup Var(u, v)) = \emptyset$, and if neither u nor l is a variable, then $Root(u) = Root(l)$. Root Rewriting may not be applied hereafter to the equation $u \approx l$. This transformation represents a *leftmost* rewrite step at the root, and avoids rewriting a variable occurrence if possible.[1]

Root Imitation: If x is a variable and $f \in \Sigma_n$ with $n > 0$, then we have

$$\{x \approx v\} \cup S \Longrightarrow \{x \approx f(y_1, \ldots, y_n), x \approx v\} \cup S, \tag{5}$$

where the $y_1, \ldots, y_n$ are *new* variables and if v is not a variable, then $f = Root(v)$. As a part of this transformation, we immediately apply Variable Elimination to the new equation $x \approx f(y_1, \ldots, y_n)$.

As in the transformations in $\mathcal{ST}$, recall that systems are multisets, and the unions above are *multiset unions*.

Thus, given a set of equations E and a system S to be E-unified, we say that E-Unify$(S) = \theta$ iff there exists a sequence of transformations from the set $\mathcal{BT}$

$$S \Longrightarrow S_1 \Longrightarrow \ldots \Longrightarrow S',$$

with S' in solved form and $\theta = \sigma_{S'}|_{Var(S)}$.

Example 5.1.3 Let $E = \{fgz \doteq z\}$ and $S = \{hx \approx hgfx\}$. Then we have the following sequence of transformations:

$$
\begin{aligned}
hx \approx hgfx &\Longrightarrow_{\text{dec}} && x \approx gfx \\
&\Longrightarrow_{\text{imit,vel}} && x \approx gy_1, \; gy_1 \approx gfgy_1 \\
&\Longrightarrow_{\text{dec}} && x \approx gy_1, \; y_1 \approx fgy_1 \\
&\Longrightarrow_{\text{rrw}} && x \approx gy_1, \; y_1 \approx z', \; fgz' \approx fgy_1 \\
&\Longrightarrow_{\text{vel}} && x \approx gy_1, \; y_1 \approx z', \; fgy_1 \approx fgy_1 \\
&\Longrightarrow_{\text{triv}} && x \approx gy_1, \; y_1 \approx z'
\end{aligned}
$$

---

[1] Strictly speaking this transformation is something like a paramodulation step at the root, except that the terms u and l are not unified. The point is that the juxtaposition of an equation between the terms u and v imitates the way a rewrite step occurs in the proof that two terms E-unify, and is not just paramodulation, since further rewrites can take place below the root of u and l.

Therefore, $E\text{-Unify}(S) = [gy_1/x] = \theta$ is an E-unifier of the terms hx and $hgfx$, as shown by the rewrite sequence

$$\theta(hx) = hgy_1 \longleftrightarrow_{[11, z' \doteq fgz', y_1/z']} hgfgy_1 = \theta(hgfx).$$

The general idea here is that given some $\theta \in U_E(S)$, we wish to show that it is always possible to find some $\sigma \in U_E(S)$ such that $\sigma \leq_E \theta[Var(S)]$; in particular, this will be accomplished if we can find a substitution $\sigma \in U_E(S)$ such that $\theta =_E \sigma \circ \theta[Var(S)]$. The basic method of the transformations is to find solved equations $x \approx t$ such that $\theta(x) \overset{*}{\longleftrightarrow}_E \theta(t)$, so that, by an argument similar to that used in lemma 3.3.4, we have $\theta =_E [t/x] \circ \theta$. The sequence of solved equations found may be thought of as 'pieces' of the substitution θ, and the set of solved equations collected constitute successive approximations of the substitution θ, namely, $\sigma_1 = [t_1/x_1], \sigma_2 = [t_1/x_1] \circ [t_2/x_2], \dots$. When we have approximated θ sufficiently to E-unify the system, we may stop.

In this context, Root Imitation represents a 'minimal approximation' of a substitution. This corresponds to case 3.b in our previous analysis of E-unification, where some rewrite steps occur, but *not* at the root, and one of the terms is a variable. We assume u is some variable x, and then either (i) v is a compound term $f(v_1, \dots, v_n)$, where $n \neq 0$, or (ii) v is a variable. In case (i), we know that $\theta'(u) = f(t_1, \dots, t_n)$ for some terms $t_1, \dots, t_n$, and $t_i \overset{*}{\longrightarrow}_E \theta'(v_i)$ for $1 \leq i \leq n$. But we can not yet tell the exact identity of the terms $t_1, \dots, t_n$; we know only that $Root(\theta(x)) = f$. Thus we assume that $\theta'(x) = f(y_1, \dots, y_n)$, where the new variables $y_1, \dots, y_n$ are "placeholders" for the rest of the binding, and will be found at some later point. Such a binding for x may be called a *general binding for x*. We may roughly think of this as extending the substitution $\theta' = [f(t_1, \dots, t_n)/x] \cup \theta''$ into a substitution

$$\hat{\theta}' = [f(y_1, \dots, y_n)/x] \circ [t_1/y_1, \dots, t_n/y_n] \cup \theta'',$$

where clearly $\hat{\theta}' = \theta'[D(\theta')]$. By solving the equation $x \approx f(y_1, \dots, y_n)$, we have found a piece of this extended substitution. The bindings for the new variables will be found later and substituted in using variable elimination. In case (ii), where both u and v are variables, we know that $\theta'(u) = f(t_1, \dots, t_n)$ and $\theta'(v) = f(t'_1, \dots, t'_n)$ for some terms $t_1, \dots, t_n$, $t'_1, \dots, t'_n$ where $t_i \overset{*}{\longleftrightarrow}_E t'_i$ for $1 \leq i \leq n$. In this case we "guess" a general binding for u, and then this case is reduced to the previous one. Thus we must guess the root symbol of the binding; this 'don't know' non-determinism clearly presents implementation problems, but for the present we are only concerned with demonstrating the completeness of a very general set of transformations; in Section §6.2 we show how this can be avoided.

One interesting special case where root imitation is applicable is in E-unifying an equation of the form $x \approx t$, where $x \in Var(t)$, i.e., when the *occur check* fails for x. Although such an equation cannot have a mgu, it is potentially E-unifiable by rewriting at the root (e.g., $[a/x] \in U_E(x, f(x))$ for $E = \{a \doteq f(a)\}$) or by rewriting below the root, as shown in Example 5.1.3 for the equation $x \approx g(f(x))$. To E-unify an equation $x \approx f(v_1, \ldots, v_n)$ where the occur check fails for x and no rewrite occurs at the root of $f(v_1, \ldots, v_n)$, we simulate rewriting below the root by the use of Root Imitation and Term Decomposition, imitating the root f with a general binding for x, and decomposing, thus distributing the occur check into at least one of the equations $y_1 \approx v_1, \ldots, y_n \approx v_n$, whereupon we may apply Root Rewriting or Root Imitation again to that equation. At some point we must find an application of Root Rewriting if we are to eliminate the occur check. Unfortunately, it is possible to create an infinite series of equations isomorphic up to renaming by repeatedly applying Root Imitation and Term Decomposition:

$$x \approx f(x) \implies_{\text{imit,dec}} x \approx f(y_1), y_1 \approx f(y_1)$$
$$\implies_{\text{imit,dec}} x \approx f(f(y_2)), y_1 \approx f(y_2), y_2 \approx f(y_2) \ldots.$$

Obviously this problem can not arise unless the occur check fails. In §6.2 we show that we can eliminate such redundant sequences without affecting the completeness of the procedure.

5.2 Soundness of the Set $\mathcal{BT}$

The following lemmas will be used to show that our procedure is sound. The first is a straightforward adaptation of Lemma 3.3.7.

Lemma 5.2.1 If $S \implies S'$ using Trivial or Variable Elimination, then $U_E(S) = U_E(S')$.

Proof. As with standard unification, the only difficulty is with Variable Elimination. We must show that $U_E(\{x \approx v\} \cup S) = U_E(\{x \approx v\} \cup \sigma(S))$ where $\sigma = [v/x]$ and $x \notin Var(v)$. For any substitution θ, if $\theta(x) \xleftrightarrow{*}_E \theta(v)$, then $\theta =_E \sigma \circ \theta$, since $\sigma \circ \theta$ differs from θ only at x, but $\theta(x) \xleftrightarrow{*}_E \theta(v) = \sigma \circ \theta(x)$. Thus,

$$\theta \in U_E(\{x \approx v\} \cup S)$$
$$\text{iff } \theta(x) \xleftrightarrow{*}_E \theta(v) \text{ and } \theta \in U_E(S)$$

$$\text{iff } \theta(x) \overset{*}{\longleftrightarrow}_E \theta(v) \text{ and } \sigma \circ \theta \in U_E(S); \text{ by lemma 4.1.4}$$

$$\text{iff } \theta(x) \overset{*}{\longleftrightarrow}_E \theta(v) \text{ and } \theta \in U_E(\sigma(S))$$

$$\text{iff } \theta \in U_E(\{x \approx v\} \cup \sigma(S)).$$

$\square$

Lemma 5.2.2 If $S \implies S'$ using one of Term Decomposition, Root Rewriting, or Root Imitation, then $U_E(S') \subseteq U_E(S)$.

Proof. The basic idea here is that these transformations do not preserve those E-unifiers which require a rewrite step or an application of root imitation, but do not introduce the possibility of new E-unifiers. There are three cases.

(i) *Term Decomposition*: If we have $\theta(s_i) \overset{*}{\longleftrightarrow}_E \theta(t_i)$, for $1 \leq i \leq n$, then $\theta(f(s_1, \ldots, s_n)) \overset{*}{\longleftrightarrow}_E \theta(f(t_1, \ldots, t_n))$, so clearly $S \implies_{\text{dec}} S'$ and $\theta \in U_E(S')$ implies that $\theta \in U_E(S)$.

(ii) *Root Rewriting*: If $\theta(u) \overset{*}{\longleftrightarrow}_E \theta(l)$, $\theta(r) \overset{*}{\longleftrightarrow}_E \theta(v)$ for some variant $l \doteq r$ of an equation from $E \cup E^{-1}$, then

$$\theta(u) \overset{*}{\longleftrightarrow}_E \theta(l) \longleftrightarrow_{[\epsilon, l \doteq r, \theta]} \theta(r) \overset{*}{\longleftrightarrow}_E \theta(v).$$

Thus $S \implies_{\text{rrw}} S'$ and $\theta \in U_E(S')$ implies that $\theta \in U_E(S)$.

(iii) *Root Imitation*: This is in two parts. First we add an equation $x \approx f(y_1, \ldots, y_n)$ to the system, and then we apply Variable Elimination. Since we showed the soundness of Variable Elimination, we simply observe that if $S \implies_{\text{imit}} S'$ then $S \subseteq S'$, so clearly $\theta \in U_E(S')$ implies that $\theta \in U_E(S)$.

In the case of Root Rewriting, the inclusion is always proper if the equation is not ground, since E-unifiers of the new system must account for the variables in the equation used in the rewrite step. The inclusion is also proper with Root Imitation, since new variables are introduced again. $\square$

Using these lemmas, we have the major result of this subsection.

Theorem 5.2.3 (Soundness) If $S \overset{*}{\implies} S'$, with S' in solved form, then $\sigma_{S'}|_{Var(S)} \in U_E(S)$.

Proof. Using the previous two lemmas and a trivial induction on the length of transformation sequences, we have that $\sigma_{S'} \in U_E(S)$. But since the restriction has no effect as regards the terms in S, we must have also that $\sigma_{S'}|_{Var(S)} \in U_E(S)$. $\square$

5.3 Completeness of the Set $\mathcal{BT}$

It is a testament to the power and elegance of the technique of unification by transforming systems of equations that it can be adapted to E-unification by adding only two additional transformations, and that this method, as we prove in this section, can non-deterministically find a $CSU_E(S)[V]$ for *arbitrary E, S, and V.*

In order to prove the completeness of the set $\mathcal{BT}$, we must show that if $\theta \in U_E(S)$, then there exists some sequence of transformations resulting in a solved form S' such that $\sigma_{S'} \leq_E \theta[Var(S)]$. The strategy we adopt is to take a representation for the fact that $\theta \in U_E(S)$, and let its structure determine the sequence of transformations. In particular, we shall proceed as follows. First, we observe that for any system $S = \{u_1 \approx v_1, \ldots, u_n \approx v_n\}$ there must exist sequences of rewrite steps $\theta(u_1) \xleftrightarrow{*}_E \theta(v_1), \ldots, \theta(u_n) \xleftrightarrow{*}_E \theta(v_n)$ proving that $\theta \in U_E(S)$, and we form an E-unifier θ' similar to the extension of θ as defined above in Section §5.1. Then we define an extension $\widehat{\theta'}$ of θ' and a system of equations $B_{\theta'}$ which account for all the potential uses of general bindings by Root Imitation used in building up parts of the substitution θ'. The next step is to show how, for every sequence of rewrite steps $\theta(u_i) \xleftrightarrow{*}_E \theta(v_i)$ there corresponds a *equational proof tree* which represents the sequence of rewrite steps in a more convenient form, and then define a *proof system* $< \widehat{\theta'}, B_{\theta'}, P >$, where P is a set of equational proof trees corresponding to all the equations in S. This proof system is essentially a 'preprocessing' of the original θ, S, and the sequences of rewrite steps showing that $\theta \in U_E(S)$, in which all the syntactic materials possibly used by the transformation rules have been collected together in a fashion which makes the completeness of the set $\mathcal{BT}$ more evident. We then define a set of proof transformation rules analogous to the set of transformations for systems which decompose the set of proof trees to a trivial form; this sequence of proof transformations corresponds in a natural way to a sequence of transformations on systems of equations which, when applied to the original system S, finds a system S' in solved form such that $\sigma_{S'} \leq_E \theta[Var(S)]$. This is the essence of the method of proving non-deterministic completeness: we show that for any $\theta \in U_E(S)$, with E and S arbitrary, there always exists *some* sequence of transformations which finds a E-unifier more general than θ.

We showed in Section §5.1 how for any $\theta \in U_E(S)$, there corresponds a set of rewrite sequences and an extension θ' of θ incorporating all the matching substitutions. We provide a more rigorous formulation of this as follows. We need one preliminary lemma.

Lemma 5.3.1 If

$$u = u_0 \longleftrightarrow_{[\alpha_1, l_1 \doteq r_1, \rho_1]} u_1 \cdots \longleftrightarrow_{[\alpha_n, l_n \doteq r_n, \rho_n]} u_n = v,$$

for some sequence of equations from $E \cup E^{-1}$, then for any σ we have

$$\sigma(u_0) \longleftrightarrow_{[\alpha_1, l_1 \doteq r_1, \rho_1 \circ \sigma]} \sigma(u_1) \cdots \longleftrightarrow_{[\alpha_n, l_n \doteq r_n, \rho_n \circ \sigma]} \sigma(u_n). \qquad (*)$$

Proof. We proceed by induction on n. If $n = 0$ then the result holds trivially. Now assume the hypothesis for all such sequences of length less than n for $n > 0$. For a sequence of length n we have

$$\sigma(u_0) \longleftrightarrow_{[\alpha_1, l_1 \doteq r_1, \rho_1 \circ \sigma]} \sigma(u_1) \cdots \longleftrightarrow_{[\alpha_{n-1}, l_{n-1} \doteq r_{n-1}, \rho_{n-1} \circ \sigma]} \sigma(u_{n-1})$$

and $u_{n-1} \longleftrightarrow_{[\alpha_n, l_n \doteq r_n, \rho_n]} u_n$, that is, $u_{n-1}/\alpha_n = \rho_n(l_n)$ and $u_n = u_{n-1}[\alpha_n \leftarrow \rho_n(r_n)]$. But then, since $\alpha_n \in Dom(u_{n-1})$ we have $\sigma(u_{n-1})/\alpha_n = \sigma(u_{n-1}/\alpha_n) = \sigma(\rho_n(l_n))$ and

$$\sigma(u_n) = \sigma(u_{n-1}[\alpha_n \leftarrow \rho_n(r_n)]) = \sigma(u_{n-1})[\alpha_n \leftarrow \sigma(\rho_n(r_n))],$$

and so therefore $\sigma(u_{n-1}) \longleftrightarrow_{[\alpha_n, l_n \doteq r_n, \rho_n \circ \sigma]} \sigma(u_n)$, from which $(*)$ follows. $\qquad \square$

Lemma 5.3.2 For any system $S = \{u_1 \approx v_1, \ldots, u_n \approx v_n\}$, if $\theta \in U_E(S)$ then there exists some idempotent $\theta' \in U_E(S)$ such that $\theta' \leq \theta[Var(S)]$ and some set of rewrite sequences $R = \{\Pi_1, \ldots, \Pi_n\}$ proving[2] that θ' E-unifies each equation in S, where each such sequence has the form

$$\theta'(u) = u_0 \longleftrightarrow_{[\alpha_1, l_1 \doteq r_1, \theta']} u_1 \cdots \longleftrightarrow_{[\alpha_m, l_m \doteq r_m, \theta']} u_m = \theta'(v). \qquad (1)$$

Proof. Let $\{\rho_1, \ldots, \rho_m\}$ be the set of all matching substitutions used in all the n rewrite sequences in R; as in the beginning of Section §5.1 we may create an extension incorporating all the matching substitutions used in a rewrite sequence, since all occurrences of equations in all rewrite sequences are assumed to be renamed away from each other and from $Var(S)$. Thus, let $\theta'' = \theta \cup \rho_1 \cup \ldots \cup \rho_m$, so that we have

$$\theta''(u) = u_0 \longleftrightarrow_{[\alpha_1, l_1 \doteq r_1, \theta'']} u_1 \cdots \longleftrightarrow_{[\alpha_m, l_m \doteq r_m, \theta'']} u_m = \theta''(v). \qquad (2)$$

Now, because all equations in R are variants, we have $\theta'' = \theta[Var(S)]$. If θ'' is not idempotent then there exists by Lemma 7.1.26 a renaming

---

[2] R is a set of *specific* sequences of rewrite steps, denoted by Π_i; see Definition 6.1.1.

substitution ρ' and an idempotent $\theta' = \theta'' \circ \rho'$ such that $\theta' \leq \theta''[W]$ where W is the set of all variables in S, in the set of variants of equations used in R, and in the support of θ'. Clearly we have $\theta' \leq \theta'' = \theta[Var(S)]$, and finally, by our preceeding lemma, we may apply the substitution ρ' to the entire sequence (2) to obtain the sequence (1). $\square$

Let us assume in what follows that such a set of rewrite sequences and such a θ' is fixed. We now proceed to define the set $B_{\theta'}$ and the extension $\widehat{\theta'}$ which account for the general bindings used by root imitation.

Definition 5.3.3 For a given substitution θ', let us define a *general expansion of* θ', denoted $\widehat{\theta'}$, and the corresponding *system of general bindings for* θ', denoted $B_{\theta'}$, as follows. For each $x \in D(\theta')$, let $\theta'_x = \theta'|_{\{x\}}$. For each such θ'_x, define inductively the substitution $\widehat{\theta'_x}$ and the set $B_{\theta'_x}$ as follows. If $\theta'_x = [t/x]$ with $|t| = 0$, i.e., t is either a constant or a variable, then let $\widehat{\theta'_x} = \theta'_x$ and $B_{\theta'_x} = \emptyset$. Otherwise, if $\theta'_x = [f(t_1, \ldots, t_n)/x]$, then for some new variables $y_1, \ldots, y_n$, let $\theta'_{y_i} = [t_i/y_i]$ for $1 \leq i \leq n$, let

$$\widehat{\theta'_x} = \theta'_x \cup \widehat{\theta'_{y_1}} \cup \ldots \cup \widehat{\theta'_{y_n}},$$

and let

$$B_{\theta'_x} = \{x \approx f(y_1, \ldots, y_n)\} \cup B_{\theta'_{y_1}} \cup \ldots \cup B_{\theta'_{y_n}}.$$

Finally, let $\widehat{\theta'} = \bigcup_{x \in D(\theta')} \widehat{\theta'_x}$ and $B_{\theta'} = \bigcup_{x \in D(\theta')} B_{\theta'_x}$.

For example, if $\theta' = [g(f(a), b)/x, z/y]$, then

$$\widehat{\theta'} = [g(f(a), b)/x, f(a)/y_1, a/y_2, b/y_3, z/y],$$

and

$$B_{\theta'} = \{x \approx g(y_1, y_3), y_1 \approx f(y_2)\}.$$

The following lemma demonstrates the essential properties of $\widehat{\theta'}$ and and $B_{\theta'}$ needed in our completeness proof.

Lemma 5.3.4 For any substitution $\theta' \in U_E(S)$ for some S, there exists some $\widehat{\theta'}$ and $B_{\theta'}$ such that
  (i) $\widehat{\theta'}$ and $B_{\theta'}$ are unique up to the choice of new variables in $D(\widehat{\theta'}) - D(\theta')$;
  (ii) θ' is idempotent iff $\widehat{\theta'}$ is idempotent;
  (iii) $\widehat{\theta'} = \theta'[D(\theta') \cup Var(S)]$, with the result that $\widehat{\theta'} \in U_E(S)$;
  (iv) $\widehat{\theta'} \in U(B_{\theta'})$.

Proof. By a simple induction on $|t|$ we can show that $\widehat{\theta'}_x$ exists for any $\theta' = [t/x]$, and so clearly $\widehat{\theta'}$ and $B_{\theta'}$ exist, and since the only place in the construction for non-uniqueness is in picking the new variables, the result is always unique up to this choice, showing (i). By an induction which follows the construction of $\widehat{\theta'}$ we can show that $I(\widehat{\theta'}) = I(\theta')$ and $D(\widehat{\theta'}) = D(\theta') \cup Y$, where Y is the set of new variables chosen. Now, since Y consists of new variables, we must have $Y \cap I(\widehat{\theta'}) = \emptyset$, so that $D(\widehat{\theta'}) \cap I(\widehat{\theta'}) = \emptyset$ iff $D(\theta') \cap I(\theta') = \emptyset$. But then by Lemma 7.1.25, we have (ii). Again, as a consequence of the set Y being new variables, (iii) must hold. Finally, note that by our definition, for any single binding t/x in $\widehat{\theta'}$, either $|t| = 0$ or t is some compound term $f(t_1, \ldots, t_n)$ such that there exists an equation $x \approx f(y_1, \ldots, y_n)$ in $B_{\theta'}$ and some bindings $t_1/y_1, \ldots, t_n/y_n$ in $\widehat{\theta'}$. Thus by a simple induction on the construction of $B_{\theta'}$ we see that (iv) holds. $\square$

The idea here is that we wish to preprocess the substitution θ' in order to determine the set of general bindings which *might* be used in a transformation by Root Imitation. Thus we determine in advance the set of equations potentially introduced by Root Imitation and also the extensions to the substitution which 'fill in' these general bindings.

Now we define our formalism for the fact that such a substitution E-unifies a pair of terms.

Definition 5.3.5 Let θ be some idempotent substitution, and let $\widehat{\theta'}$ and $B_{\theta'}$ be as above. The set of *proof trees associated with* $\widehat{\theta'}$ is defined inductively as follows. For simplicity we use $*$ as a syntactic variable for one of the symbols $\asymp$, $\sim$, or $=$.

(i) (Axioms) For every term u, the one node tree labeled with $u = u$ is a proof tree associated with $\widehat{\theta'}$. For every two terms $u \neq v$, at least one of which is a variable and the other a constant or a variable, such that $\widehat{\theta'}(u) = \widehat{\theta'}(v)$, the one node tree labeled with $u = v$ is a proof tree associated with $\widehat{\theta'}$. Thus, axioms are trivial proofs that identical terms are E-unifiable or that a variable in the domain of the substitution associated with the proof trivially E-unifies with some term. Note that in the latter case, the axiom will be formed from two terms x and t, where $x \notin Var(t)$, and that it is not necessary that $\widehat{\theta'}(x) = t$.

(ii) (Term Decomposition) Let u and v be an pair of terms, $f \in \Sigma_n$, and $u_1, \ldots, u_n, v_1, \ldots, v_n$ be terms such that
    (a) If u is a variable, then $\widehat{\theta'}(u) = f(u_1, \ldots, u_n)$, otherwise $u = f(u_1, \ldots, u_n)$, and
    (b) If v is a variable, then $\widehat{\theta'}(v) = f(v_1, \ldots, v_n)$, otherwise $v = f(v_1, \ldots, v_n)$.

Given any n proof trees $T_1, \ldots, T_n$ associated with $\widehat{\theta'}$, where each T_i is a proof tree whose root is labeled with $u_i * v_i$, the tree T whose root is labeled with $u \sim v$ and such that $T/i = T_i$ for $1 \leq i \leq n$ is a proof tree associated with $\widehat{\theta'}$. Thus, a proof tree whose root is labeled with $u \sim v$ represents the fact that $\widehat{\theta'}(u) \overset{*}{\longleftrightarrow}_E \widehat{\theta'}(v)$, where no rewrite steps occur at the root. Note that if either of the terms u or v is a variable, then we must instantiate it before decomposing it in the proof tree; if a term is compound it is simply decomposed, without the substitution being applied.

(iii) (Root Rewriting) Let u and v be an pair of terms and $l_i \doteq r_i$ for $1 \leq i \leq m$ be variants of equations from $E \cup E^{-1}$. Furthermore, let $T_1, \ldots, T_{m+1}$ be proof trees associated with $\widehat{\theta'}$, where T_1 is a proof tree whose root is labeled with either $u = l_1$ or $u \sim l_1$, and for $2 \leq i \leq m$, T_i is a proof tree whose root is labeled with either $r_{i-1} = l_i$ or $r_{i-1} \sim l_i$, and T_{m+1} is a proof tree whose root is labeled with either $r_m = v$ or $r_m \sim v$. Then the tree T whose root is labeled with $u \asymp v$ and such that $T/i = T_i$ for $1 \leq i \leq m+1$ is a proof tree associated with $\widehat{\theta'}$. This shows the effect of all the rewrites occurring at the root in $\widehat{\theta'}(u) \overset{*}{\longleftrightarrow}_E \widehat{\theta'}(v)$.

In general, we regard the nodes of a proof tree as *unordered* pairs of terms, in accordance with the unordered nature of equations. A proof tree associated with $\widehat{\theta'}$ whose root is labeled with $u * v$ will be denoted by the pair $\langle \widehat{\theta'}, (u * v) \rangle$, or simply $(u * v)$ if the substitution is available from context.[3] It should be obvious that with any set of proof trees P we may associate a system of equations S, namely, the set of equations occurring in the roots of the proof trees in the set P; this is called the *root system* of P.

Finally, a triple $\langle \widehat{\theta'}, B_{\theta'}, P \rangle$ is a *proof system for θ and S* if θ' is an idempotent substitution incorporating all the matching substitutions used in some particular sequences of rewrite steps showing that $\theta \in U_E(S)$, as in Lemma 5.3.2, if $\widehat{\theta'}$ is the general expansion of θ' and $B_{\theta'}$ is the set of general bindings for θ', and finally if P is a set of proof trees associated with $\widehat{\theta'}$ with a root system S. Note that as a consequence of these definitions, for each subproof $(x \sim v)$ occurring somewhere in a proof in P, there exists some equation $x \approx t$ in $B_{\theta'}$; this corresponds to the equation possibly added to the system by some application of Root Imitation to the equation $x \approx v$.

We shall prove that these proof systems are sound and complete with respect to the definition of E-unification after presenting an illustration based on a variation of Example 5.1.3.

---

[3] Note carefully that $u * v$ is the *label* of a proof tree node, and $(u * v)$ is a proof tree whose root node is labeled with $u * v$.

Example 5.3.6 Let $E = \{ f(g(z)) \doteq z \}$. The rewrite sequence which proves that $\theta = [g(a)/x]$ is an E-unifier of $S = \{h(x) \approx h(g(f(x)))\}$ is

$$\theta(h(x)) = h(g(a)) \;\longrightarrow_{[11,\, z' \doteq f(g(z')),\, a/z']}\; h(g(f(g(a)))) = \theta(h(g(f(x)))),$$

and so we may form the E-unifier $\theta' = [g(a)/x, a/z']$ and then the general expansion $\widehat{\theta}' = [g(a)/x, a/y_1, a/z']$ and the set of general bindings $B_{\theta'} = \{x \approx g(y_1)\}$. The proof system for θ and S is thus $\langle \widehat{\theta}', B_{\theta'}, P \rangle$, where P is the set consisting of the single proof tree

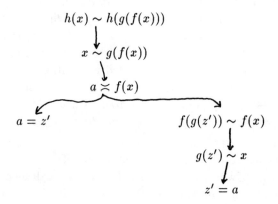

The root system of P is $\{h(x) \approx h(g(f(x)))\}$. (Compare with Example 5.1.3.)

When convenient, we shall represent the (partial) structure of a proof tree with root node $u * v$ and subtrees $P_1, \ldots, P_n$ in the prefix form $u * v[P_1, \ldots, P_n]$, e.g., variously representing the subtree with root node $a \asymp f(x)$ above in any of the forms

$$(a \asymp f(x)), \qquad a \asymp f(x)[a = z', (f(g(z')) \sim f(x))],$$

or

$$a \asymp f(x)[a = z', f(g(z')) \sim f(x)[g(z') \sim x[z' = a]]].$$

This linear notation will make it somewhat easier to manipulate proof trees.

Our next two theorems show that our proof representation is sound and complete with respect to the definition of E-unification.

Theorem 5.3.7 For some given substitution θ, system S, and set of equations E, if $\langle \widehat{\theta}', B_{\theta'}, P \rangle$ is a proof system for θ and S, then $\theta \in U_E(S)$.

Proof. By Lemmas 5.3.2 and 5.3.4, we have $\widehat{\theta'} = \theta' \leq \theta[Var(S)]$, and so if we can show that for each proof tree $(u*v)$ in P, we have $\widehat{\theta'}(u) \overset{*}{\longleftrightarrow}_E \widehat{\theta'}(v)$, then by Corollary 4.1.5 we shall have our result. Thus let $T = (u*v)$ be an arbitrary proof tree in P. We proceed by induction on the number n of tree nodes in T. If $n = 1$, then $\widehat{\theta'}(u) = \widehat{\theta'}(v)$ by definition. Now assume that the result holds for all proof trees with less than n nodes, with $n > 1$, and suppose T contains n nodes. There are two cases.

(i) If the root node of T is labeled with $u \sim v$, then as above we suppose f is the root of $\widehat{\theta'}(u)$ and let $u_1, \ldots, u_m, v_1, \ldots, v_m$ be terms such that

(a) If u is a variable, then $\widehat{\theta'}(u) = f(u_1, \ldots, u_m)$, otherwise $u = f(u_1, \ldots, u_m)$,

(b) If v is a variable, then $\widehat{\theta'}(v) = f(v_1, \ldots, v_m)$, otherwise $v = f(v_1, \ldots, v_m)$.

There are thus proof trees

$$T/1 = (u_1 * v_1), \ldots, T/m = (u_m * v_m)$$

and by the hypothesis, $\widehat{\theta'}(u_i) \overset{*}{\longleftrightarrow}_E \widehat{\theta'}(v_i)$ for $1 \leq i \leq m$. By changing the rewrite addresses $\alpha_1, \alpha_2, \ldots$ in the i_{th} such sequence to $i\alpha_1, i\alpha_2, \ldots$, and concatenating these m new rewrite sequences, we see that $\widehat{\theta'}(u) \overset{*}{\longleftrightarrow}_E \widehat{\theta'}(v)$. (Note how the idempotency of $\widehat{\theta'}$ is used here.)

(ii) If the root node of T is labeled with $u \asymp v$ then there are proof trees

$$T/1 = (u * l_1), \ T/2 = (r_1 * l_2), \ldots, T/k + 1 = (r_k * v),$$

where the $l_i \doteq r_i$ are variants of equations from $E \cup E^{-1}$, and, by hypothesis,

$$\widehat{\theta'}(u) \overset{*}{\longleftrightarrow}_E \widehat{\theta'}(l_1), \ldots, \widehat{\theta'}(r_k) \overset{*}{\longleftrightarrow}_E \widehat{\theta'}(v),$$

and so

$$\widehat{\theta'}(u) \overset{*}{\longleftrightarrow}_E \widehat{\theta'}(l_1) \overset{}{\longleftrightarrow}_{[\epsilon, l_1 \doteq r_1, \widehat{\theta'}]} \widehat{\theta'}(r_1) \overset{*}{\longleftrightarrow}_E \ldots \widehat{\theta'}(r_k) \overset{*}{\longleftrightarrow}_E \widehat{\theta'}(v),$$

with the result that again $\widehat{\theta'}(u) \overset{*}{\longleftrightarrow}_E \widehat{\theta'}(v)$. □

Theorem 5.3.8 If $\theta \in U_E(S)$, then there exists a proof system $\langle \widehat{\theta'}, B_{\theta'}, P \rangle$ associated with θ and S.

Proof. As shown in Lemma 5.3.2, if $\theta \in U_E(S)$ then there must exist some particular sequence of rewrites proving this fact, and an idempotent E-unifier θ' incorporating all the matching substitutions used in rewrite steps. Then by Lemma 5.3.4 we know that $\widehat{\theta'}$ and $B_{\theta'}$ must exist, so if we can show that for any $u \approx v \in S$ there exists an equational proof tree $(u * v)$ associated with $\widehat{\theta}'$, then we can simply collect all these trees together to form P and we have our result.

Thus we shall show by induction that for any particular sequence

$$\widehat{\theta}'(u) = u_0 \xmapsto{\hspace{0.5em}}_{[\alpha_1, l_1 \doteq r_1, \widehat{\theta}']} u_1 \cdots \xmapsto{\hspace{0.5em}}_{[\alpha_n, l_n \doteq r_n, \widehat{\theta}']} u_n = \widehat{\theta}'(v),$$

we have a proof tree $(u * v)$ associated with $\widehat{\theta}'$. With any such rewrite sequence, we associate a complexity measure

$$\mu = \{|u_0|, |u_1|, \ldots, |u_n|\},$$

that is, a multiset of the depths of the terms $u_0, \ldots, u_n$. Our proof proceeds by induction on μ, using the standard multiset ordering.

Basis. $\mu = \{k\}$ and either $u = v$ or one of u, v is a variable. Then by Definition 5.3.5 $(u = v)$ is a proof tree associated with $\widehat{\theta}'$. (This constitutes a sufficient basis since it includes the case $k = 0$.)

Induction. Assume there exists a corresponding proof tree for all such rewrite sequences with complexity strictly less than μ, and consider a sequence with complexity μ, as above. There are three cases.

(i) $\mu = \{k\}$ where $u \neq v$ and neither of u, v is a variable. Now we must have $Root(u) = Root(v)$, and since $u \neq v$, both are compound terms, i.e., $k > 0$. Thus $\widehat{\theta}'(u) = u_0 = \widehat{\theta}'(v)$ and $u = f(s_1, \ldots, s_m)$ and $v = f(t_1, \ldots, t_m)$ for some terms $s_1, \ldots, s_m, t_1, \ldots, t_m$. Then $\widehat{\theta}'(s_i) = u_0/i = \widehat{\theta}'(t_i)$ with $|u_0/i| < |u_0|$ for $1 \leq i \leq m$, and by hypothesis, there are proof trees $(s_1 * t_1), \ldots, (s_m * t_m)$ associated with $\widehat{\theta}'$, and so by definition there must exist a proof tree $u \sim v[(s_1 * t_1), \ldots, (s_m * t_m)]$ associated with $\widehat{\theta}'$. (This proof tree will naturally contain no rewrite nodes.)

(ii) $\mu = \{k_0, k_1, \ldots, k_n\}$ for $n > 0$, and there is no rewrite at the root of any u_i. In this case, $Root(\widehat{\theta}'(u)) = Root(\widehat{\theta}'(v))$, and the subterms are pairwise E-congruent. More precisely, let $f = Root(\widehat{\theta}'(u))$ be a function symbol of arity m, and $s_1, \ldots, s_m, t_1, \ldots, t_m$ be terms such that

    (a) If u is a variable, then $\widehat{\theta}'(u) = f(s_1, \ldots, s_m)$, otherwise $u = f(s_1, \ldots, s_m)$, and

    (b) If v is a variable, then $\widehat{\theta}'(v) = f(t_1, \ldots, t_m)$, otherwise $v = f(t_1, \ldots, t_m)$.

Then for each $1 \leq i \leq m$ we have that

$$\widehat{\theta}'(s_i) = u_0/i \xmapsto{\hspace{0.5em}}_E u_1/i \xmapsto{\hspace{0.5em}}_E \cdots \xmapsto{\hspace{0.5em}}_E u_n/i = \widehat{\theta}'(t_i),$$

with a complexity strictly less than μ. By the induction hypothesis, there exist proof trees $(s_1 * t_1), \ldots, (s_m * t_m)$ associated with $\widehat{\theta}'$, and thus by definition a proof tree

$$u \sim v[(s_1 * t_1), \ldots, (s_m * t_m)]$$

associated with $\widehat{\theta'}$. (Note that the idempotency of $\widehat{\theta'}$ is necessary in case one of u, v is a variable.)

(iii) $\mu = \{k_0, k_1, \ldots, k_n\}$ for $n > 0$, and there is a rewrite at the root of some u_i. Then we may represent the sequence as

$$\widehat{\theta'}(u) \overset{*}{\longleftrightarrow}_E \widehat{\theta'}(l_1') \longleftrightarrow_{[\epsilon, l_1' \doteq r_1', \widehat{\theta'}]} \widehat{\theta'}(r_1') \overset{*}{\longleftrightarrow}_E$$

$$\widehat{\theta'}(l_2') \longleftrightarrow_{[\epsilon, l_2' \doteq r_2', \widehat{\theta'}]} \widehat{\theta'}(r_2') \overset{*}{\longleftrightarrow}_E$$

$$\ldots$$

$$\widehat{\theta'}(l_p') \longleftrightarrow_{[\epsilon, l_p' \doteq r_p', \widehat{\theta'}]} \widehat{\theta'}(r_p') \overset{*}{\longleftrightarrow}_E \widehat{\theta'}(v)$$

for some subset $\{l_1' \doteq r_1', \ldots, l_p' \doteq r_p'\}$ of the equations used in the original sequence. But then the complexity of each of the sequences

$$\widehat{\theta'}(u) \overset{*}{\longleftrightarrow}_E \widehat{\theta'}(l_1'), \quad \widehat{\theta'}(r_1') \overset{*}{\longleftrightarrow}_E \widehat{\theta'}(l_2'), \quad \ldots, \quad \widehat{\theta'}(r_p') \overset{*}{\longleftrightarrow}_E \widehat{\theta'}(v)$$

is strictly less than μ, and by hypothesis, there are proof trees $(u * l_1')$, $(r_1' * l_2'), \ldots, (r_p' * v)$ associated with $\widehat{\theta'}$. Finally, by definition there must exist a proof tree

$$u \asymp v[(u * l_1'), \ldots, (r_p' * v)]$$

associated with $\widehat{\theta'}$. $\square$

One interesting point about this completeness proof is that it gives us a canonical way of constructing a proof tree for any particular sequence of rewrite steps proving that two terms are E-unifiable by the substitution $\widehat{\theta'}$. This is particularly useful in eliminating variables by applying substitutions to proof trees.

Lemma 5.3.9 If x is a variable, t a term, and $\widehat{\theta'}$ an idempotent general expansion such that $\widehat{\theta'}(x) = \widehat{\theta'}(t)$, and if u and v are two arbitrary terms, then there exists a proof tree $(u * v)$ associated with $\widehat{\theta'}$ iff there exists a proof tree $(u[t/x] * v[t/x])$ associated with $\widehat{\theta'}$. Furthermore, if such proof trees exist, there always exist two with the same number of $\asymp$-nodes.

Proof. Since $\widehat{\theta'}(x) = \widehat{\theta'}(t)$ we must have $\widehat{\theta'} = [t/x] \circ \widehat{\theta'}$, so by Lemma 4.1.4 we have

$$\widehat{\theta'}(u) \overset{*}{\longleftrightarrow}_E \widehat{\theta'}(v) \text{ iff } [t/x] \circ \widehat{\theta'}(u) \overset{*}{\longleftrightarrow}_E [t/x] \circ \widehat{\theta'}(v)$$
$$\text{iff } \widehat{\theta'}(u[t/x]) \overset{*}{\longleftrightarrow}_E \widehat{\theta'}(v[t/x]),$$

and so, by our previous two results, there exists a proof tree $(u * v)$ associated with $\widehat{\theta'}$ iff there exists a proof tree $(u[t/x] * v[t/x])$ associated with

$\widehat{\theta'}$. Now by structural induction, it is easy to show that for any *particular sequence* of m rewrite steps we have

$$\widehat{\theta'}(u[t/x]) \longleftrightarrow_{[\alpha_1, l_1 \doteq r_1, \rho_1]} u_1$$
$$\longleftrightarrow_{[\alpha_2, l_2 \doteq r_2, \rho_2]} u_2 \cdots$$
$$\longleftrightarrow_{[\alpha_m, l_m \doteq r_m, \rho_m]} \widehat{\theta'}(v[t/x])$$

if and only if

$$\widehat{\theta'}(u) \longleftrightarrow_{[\alpha_1, l_1 \doteq r_1, \rho_1]} u_1$$
$$\longleftrightarrow_{[\alpha_2, l_2 \doteq r_2, \rho_2]} u_2 \cdots$$
$$\longleftrightarrow_{[\alpha_m, l_m \doteq r_m, \rho_m]} \widehat{\theta'}(v).$$

But then by multiset induction on this sequence, following the proof of Theorem 5.3.8, it is easy to show that if such terms are E-congruent using this particular sequence, then proof trees exist for each equation, and that the creation of $\times$-nodes corresponds to the structure of this particular sequence, and hence the number of such nodes is the same in both trees.   □

We remark that, depending on the set E, there may exist many equivalent sequences of rewrite steps, so that we can not enforce that the number of $\times$-nodes *always* be the same for any two trees; we simply prove that there always exist two such similar trees. Also, note that it would be possible to be more precise about the structural similarity of trees created canonically from the same rewrite sequence, in the sense that their $\times$-nodes occur in the same tree addresses, but this formality is unnecessary for our purposes, so we omit it. Finally, we remark that it would not in general be possible to define a similar lemma for the case of two terms x and t such that $\widehat{\theta'}(x) \longleftrightarrow^*_E \widehat{\theta'}(t)$ without extending the substitution $\widehat{\theta'}$. The reason is that we can not use the same rewrite sequence $\longleftrightarrow^*_E$ in both cases, since there may be more rewrite steps in one than the other, and since the rewrites between $\widehat{\theta}(x)$ and $\widehat{\theta'}(t)$ may be used many times, by our assumption that all rewrite sequences contain distinct variants of equations, these would be additional instances of equations, and the extension θ' would no longer be sufficient. This problem turns out to have serious consequences in proving the completeness of the strategy of eager variable elimination (see Section §6.7).

Now we show that the transformations on systems $\mathcal{BT}$ correspond to a certain set of transformations on proof systems.

Definition 5.3.10 Let P' be a set of proof trees (possibly empty). We have the following five proof transformations.

$$\{(u * u)\} \cup P' \implies P' \tag{A}$$

$$\{u \sim v[T_1, \ldots, T_n]\} \cup P' \implies \{T_1, \ldots, T_n\} \cup P', \tag{B}$$

where u and v are both compound terms.

$$\{(x * t)\} \cup P' \implies \{(x = t)\} \cup P'[t/x], \tag{C}$$

where there are no $\asymp$-nodes in the tree $(x * t)$ (i.e., no rewrite steps), x occurs in some tree in P' and where $P'[t/x]$ denotes the result of replacing each proof tree $(u * v)$ in P by a proof tree $(u[t/x] * v[t/x])$ (the existence of such a proof tree was shown in the previous lemma).

$$\{u \asymp v[T_1, \ldots, T_n]\} \cup P' \implies \{T_1, \ldots, T_n\} \cup P' \tag{D}$$

$$\{(x \sim v)\} \cup P' \implies \{(x = t), (x \sim v)\} \cup P', \tag{E}$$

where $x \approx t \in B_{\theta'}$ and where transformation (C) is immediately applied to the axiom $(x = t)$.

These proof transformations are extended from trees to systems, so that we say $\langle \widehat{\theta'}, B_{\theta'}, P \rangle \implies \langle \widehat{\theta'}, B_{\theta'}, P' \rangle$ iff $P \implies P'$.

It should be obvious that we have taken pains to define these proof transformations by analogy with our transformations on term systems. In particular, for some proof trees P and P' with root systems S and S' respectively, if $P \implies P'$ using proof transformations (A), (B), (C), or (E), then there is a corresponding transformation on the root system $S \implies S'$ using Trivial , Term Decomposition, Variable Elimination, or Root Imitation respectively. Similarly, if $P \implies_{(D)} P'$, then we have a sequence $S \overset{*}{\implies}_{\mathrm{rrw}} S'$, with one transformation step for each rewrite step left to right in the proof tree transformed in P.

Now we may prove the correctness of these proof transformations, after which we shall give an example of their use.

Lemma 5.3.11 If $\langle \widehat{\theta'}, B_{\theta'}, P \rangle$ is a proof system and $P \implies P'$ using one of the transformations (A)–(E), then $\langle \widehat{\theta'}, B_{\theta'}, P' \rangle$ is a proof system.

Proof. Clearly, the only point at issue is whether the new set P' is a set of proof trees associated with $\widehat{\theta'}$. In case (A), P' differs from P only in having one less proof tree, so clearly if P is a set of proof trees associated with $\widehat{\theta'}$, so is P'. In the case of transformations (B) and (D), since proof trees were defined inductively, for any proof tree T associated with $\widehat{\theta'}$, where T is not an axiom, the subtrees $T/1, \ldots, T/n$ for some n must still be proof trees associated with $\widehat{\theta'}$, and thus the result of either of these transformations must still be a set of proof trees associated with $\widehat{\theta'}$. If $P \implies_{(C)} P'$, then since no rewrites occur in $(x * t)$, we must have $\widehat{\theta'}(x) = \widehat{\theta'}(t)$, and so $(x = t)$ is a proof tree associated with $\widehat{\theta'}$, and by Lemma 5.3.9, there exists a proof tree $(u[t/x] * v[t/x])$ associated with $\widehat{\theta'}$. Finally, if $P \implies_{(E)} P'$, then we have simply converted an equation $x \approx t$ from $B_{\theta'}$ into a proof tree $(x = t)$, and since, by Lemma 5.3.4, $\widehat{\theta'}(x) = \widehat{\theta'}(t)$, this is an axiom tree associated with $\widehat{\theta'}$. But then $\{(x = t)\} \cup P$ is a set of proof trees associated with $\widehat{\theta'}$, and we have already shown that the subsequent application of (C) is correct. $\square$

Example 5.3.12 The transformations on the single proof tree in the proof system from Example 5.3.6 corresponding to the transformations in Example 5.1.3 are as follows.

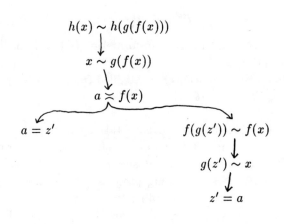

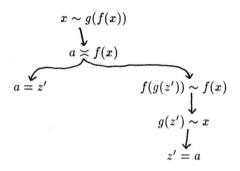

$$x \sim g(f(x))$$
$$\downarrow$$
$$a \stackrel{\sim}{\succ} f(x)$$

$$a = z' \qquad\qquad f(g(z')) \sim f(x)$$
$$\downarrow$$
$$g(z') \sim x$$
$$\downarrow$$
$$z' = a$$

$$\Downarrow_{(E)}$$

$$x = g(y_1) \qquad g(y_1) \sim g(f(g(y_1)))$$
$$\downarrow$$
$$y_1 \stackrel{\sim}{\succ} f(g(y_1))$$

$$y_1 = z' \qquad\qquad f(g(z')) \sim f(g(y_1))$$
$$\downarrow$$
$$g(z') \sim g(y_1)$$
$$\downarrow$$
$$z' = y_1$$

$$\Downarrow_{(B)}$$

$$x = g(y_1) \qquad y_1 \stackrel{\sim}{\succ} f(g(y_1))$$

$$y_1 = z' \qquad\qquad f(g(z')) \sim f(g(y_1))$$
$$\downarrow$$
$$g(z') \sim g(y_1)$$
$$\downarrow$$
$$z' = y_1$$

$$\Downarrow_{(D)}$$

$$x = g(y_1) \quad y_1 = z' \quad f(g(z')) \sim f(g(y_1))$$
$$g(z') \sim g(y_1)$$
$$z' = y_1$$

$$\Downarrow_{(C)}$$

$$x = g(y_1) \quad y_1 = z' \quad f(g(y_1)) = f(g(y_1))$$

$$\Downarrow_{(A)}$$

$$x = g(y_1) \quad y_1 = z'$$

Note that this corresponds to the solved system $S' = \{x \approx g(y_1), y_1 \approx z'\}$ found in Example 5.1.3, and that for $\theta = [g(a)/x]$ as in Example 5.3.6 we have $\sigma_{S'} \leq \theta[Var(S)]$. Our next result formalizes this by showing that the proof transformations always result in trivial proofs corresponding to solved form systems.

Lemma 5.3.13 Let $\langle \widehat{\theta'}, B_{\theta'}, P \rangle$ be a proof system. Then any sequence of proof transformations

$$P = P_0 \implies P_1 \implies \dots$$

must terminate in a system $P' = \{(x_1 = t_1), \dots, (x_n = t_n)\}$ associated with $\widehat{\theta'}$ where no transformation applies, and the root system of P' is a system in solved form.

Proof. First we show that every sequence of proof transformations must terminate. Let us define a measure of complexity for a set P of proof trees as $\mu(P) = \langle n, m \rangle$, where n is the number of variables in $D(\widehat{\theta'})$ which are not solved in the root system of P, and m is the number of nodes in all the proof trees in P. Then the lexicographic ordering on $\langle n, m \rangle$ is well-founded, and each proof transformation produces a new proof system whose measure is strictly smaller under this ordering: (A), (B), and (D) must decrease m and can not increase n; and (C) and (E) must decrease n.

Therefore the relation $\implies$ on proof systems is well-founded, and there must exist some sequence $P \overset{*}{\implies} P'$ where no transformation applies to

P'. But then P' must consist solely of axioms of the form $(x_i = t_i)$ with x_i not identical with t_i, since otherwise either (A), (B), (D), or (E) would apply, no x_i occurs in a t_i, since the two are unifiable, and furthermore each variable x_i may not occur elsewhere in the proof system, or else (C) would apply. Clearly the root system $\{x_1 \approx t_1, \ldots, x_n \approx t_n\}$ is a system in solved form.

By a simple induction on the length of the proof transformation sequence, and using Lemma 5.3.11 in the induction step, we see that P' is a proof system associated with $\widehat{\theta}'$. $\square$

Now we are ready to state the major result of this section. The completeness of our method is shown in the following theorem.

Theorem 5.3.14 (Completeness) For every $\theta \in U_E(S)$, there exists a sequence of transformations $S \stackrel{*}{\Longrightarrow} S'$ such that S' is in solved form, and $\sigma_{S'} \leq \theta[Var(S)]$.

Proof. Suppose $\theta \in U_E(S)$. Then by Theorem 5.3.8 there must exist an equational proof system $\langle \widehat{\theta}', B_{\theta'}, P \rangle$, where by Lemmas 5.3.2 and 5.3.4, we have $\widehat{\theta}' = \theta' \leq \theta[Var(S)]$. By Lemma 5.3.13 we see that there must exist some sequence of proof transformations $P \stackrel{*}{\Longrightarrow} P'$ with $P' = \{(x_1 = t_1), \ldots, (x_k = t_k)\}$ a set of proof trees associated with $\widehat{\theta}'$ to which no transformation applies, and whose root system S' is a system in solved form. By a simple induction on the length of the proof transformation sequence, we may show that there is a corresponding sequence of transformations on the root system $S \stackrel{*}{\Longrightarrow} S'$ with $S' = \{x_1 \approx t_1, \ldots, x_k \approx t_k\}$ in solved form, and since P' is a set of proof trees associated with $\widehat{\theta}'$, we have $\widehat{\theta}' \in U(S')$, so that by Lemma 3.3.4 we see that $\sigma_{S'} \leq \widehat{\theta}'$, with the result that $\sigma_{S'} \leq \widehat{\theta}' = \theta' \leq \theta[Var(S)]$. $\square$

By the soundness of the transformations, clearly any such $\sigma_{S'} \in U_E(S)$. Note that this theorem implies that $\sigma_{S'} \leq_E \theta[Var(S)]$, but is in fact a stronger result. The reason that we find more general substitutions under $\leq$ and not just $\leq_E$ is that we only perform a generalization step at the last stage, when we take the mgu of a solved form.

We may characterize the set of substitutions non-deterministically found by the set of transformations $\mathcal{BT}$ as follows.

Theorem 5.3.15 For any system S and any set of equations E, the set

$$\{\sigma_{S'}|_{Var(S)} \mid S \stackrel{*}{\Longrightarrow} S', \text{ and } S' \text{ is in solved form}\}$$

is a $CSU_E(S)$. By application of the appropriate renaming substitutions away from V, this set is a $CSU_E(S)[V]$ for any V.

Proof. We must simply verify the conditions in Definition 4.1.6. Coherence was shown in Theorem 5.2.3 and our previous result demonstrated completeness. By restricting the idempotent substitution $\sigma_{S'}$ to $Var(S)$ we satisfy purity for V empty. If V is not empty, we may suitably rename the variables introduced by each of the substitutions $\sigma_{S'}$ away from V, as shown in Lemma 3.3.11.   □

Using these results, it would be possible to implement a general procedure for E-unification in arbitrary theories by using a complete search strategy over all possible transformation sequences. In [53], a pseudo-code procedure based on Robinson's original algorithm for standard unification [139] is given for a different set of transformations for E-unification, using depth-first iterative deepening to simulate breadth-first search without excessive storage overhead. However, basing such a method on the set $\mathcal{BT}$ would be very inefficient, due to the possibility of rewriting variables in Root Rewriting. This creates many extraneous rewrite sequences, since *any* rule can unify with a variable. In addition, we must guess general bindings in the two variable case in Root Imitation to uncover potential rewrites below such equations, and, finally, we admit the potential for infinite recursion in the same rule, as remarked in Section §5.1. In the next chapter we present a new set of transformations which rectify these problems, and a proof of its completeness.

CHAPTER 6

AN IMPROVED SET OF TRANSFORMATIONS

In the last chapter we presented a set of transformations $\mathcal{BT}$ complete for arbitrary equational theories E, but which were prohibitively inefficient. In this chapter we present a restricted version of $\mathcal{BT}$, called $\mathcal{T}$, which solves these problems, and prove its completeness.

6.1 Ground Church-Rosser Systems

In this section, we shall develop techniques that will allow us to overcome the problem of possible nonterminating sequences of applications of Root Imitation. The key point is that if the equations in E were orientable and formed a canonical system R, then we could work with normalized substitutions, that is, substitutions such that $\theta(x)$ is irreducible for every $x \in D(\theta)$. If R is canonical, for every equation $x \approx v$ where x is a variable, there is a proof of the form $\theta(v) \xrightarrow{*}_R w \xleftarrow{*}_R \theta(x)$ for some irreducible w, and if θ is normalized, then the proof is in fact of the form $\theta(v) \xrightarrow{*}_R \theta(x)$, where every rule $\rho(l) \to \rho(r)$ used in this sequence applies at some *nonvariable* address β in v. Hence, for any rule in this sequence applied at a topmost level, $\theta(v/\beta)$ and $\rho(l)$ must be E-congruent. This is the motivation for a new rule, called *Lazy Paramodulation*, to replace Root Rewriting and Root Imitation:

$$\{u \approx v\} \cup S \Longrightarrow \{u/\beta \approx l, u[\beta \leftarrow r] \approx v\} \cup S, \qquad (4a)$$

where β is a nonvariable occurrence in u. A formal definition of this transformation will be given in Section 6.2, and the set of transformations $\mathcal{T}$ obtained by adding this new rule to $\mathcal{ST}$ will be given in Definition 6.2.1.

However, not every set of equations is equivalent to a canonical system of rewrite rules, and even if it is orientable with respect to some reduction ordering (thus forming a noetherian set of rules), it may not be confluent. Three crucial observations allow us to overcome these difficulties:

(1) There is no loss of generality in considering only ground substitutions;

(2) There are reduction orderings $\succ$ that are total on ground terms;

(3) Ground confluence (or equivalently, being ground Church-Rosser) is all that is needed.

These ingredients make possible the existence of *unfailing completion* procedures (Bachmair, Dershowitz, Hsiang, and Plaisted [8, 12, 9]). The main trick is that one can use *orientable ground instances of equations*, that is, ground equations of the form $\rho(l) \doteq \rho(r)$ with $\rho(l) \succ \rho(r)$, where $l \doteq r$ is a variant of an equation in $E \cup E^{-1}$. Even if $l \doteq r$ is not orientable, $\rho(l) \doteq \rho(r)$ always is if $\succ$ is total on ground terms. The last ingredient is that given a set E of equations and a reduction ordering $\succ$ total on ground terms, we can show that E can be extended to a set E^{ω} equivalent to E such that the set $R(E^{\omega})$ of orientable instances of E^{ω} is ground Church-Rosser. Furthermore, E^{ω} is obtained from E by computing critical pairs (in a hereditary fashion), treating the equations in E as two-way rules.[1]

Our "plan of attack" for the completeness proof of the new set of transformations T (given in Definition 6.2.1) is the following.

(1) Show the existence of the ground Church-Rosser completion E^{ω} of E (Theorem 6.1.7).

(2) Under the assumption that E is ground Church-Rosser, show how to extract a sequence of transformations from a rewrite sequence $\theta(s) \xrightarrow{*}_E w \xleftarrow{*}_E \theta(t)$ which demonstrates that θ is an E-unifier of s and t.

(3) For an arbitrary consistent E, show that the T-transformations are complete using Theorem 6.1.7 and a lemma which shows that the computation of critical pairs can be simulated by Lazy Paramodulation.

In (2), we shall also show that given any E-unifier θ, there is another normalized E-unifier σ such that $\sigma =_E \theta$.

Since whenever a set of equations E is inconsistent, then the identity substitution is a most general E-unifier of any pair of terms, we shall in what follows assume that the set of equations is consistent. This will have a useful consequence at a later stage of the completeness proof for this section.

It is actually more general (and more flexible) but no more complicated to deal with pairs (E, R) where E is a set of equations and R a set of rewrite rules contained in some given reduction ordering $\succ$. The set E represents

---

[1] Although a consequence of the existence of fair unfailing completion procedures proved by Bachmair, Dershowitz, Hsiang, and Plaisted [8, 12, 9], this result can be proved more directly and with less machinery.

the nonorientable part (w.r.t. $\succ$) of the system. Thus, as in Bachmair, Dershowitz, Hsiang, and Plaisted [8, 12, 9], we present our results for such systems. First, we generalize the notion of equational proof. Given a set E of equations and a rewrite system R, we define the notion of proof and rewrite proof for the pair (E, R).

Definition 6.1.1 Let E be a set of equations and R a rewrite system. For any two terms u, v, a *proof step* from u to v is a tuple $\langle u, \alpha, l, r, \sigma, v \rangle$, where α is a tree address in u, σ is a substitution, and either

$$u \longleftrightarrow_{[\alpha, l \doteq r, \sigma]} v$$

where $l \doteq r$ is a variant of an equation in $E \cup E^{-1}$, or

$$u \longrightarrow_{[\alpha, l \to r, \sigma]} v$$

where $l \to r$ is a variant of a rewrite rule in R, or

$$v \longrightarrow_{[\alpha, l \to r, \sigma]} u$$

where $l \to r$ is a variant of a rewrite rule in R.

A proof step may be (partially) described as either an *equality step* $u \longleftrightarrow_E v$, or a *rewrite step* $u \longrightarrow_R v$ or $u \longleftarrow_R v$. A *proof* that $u \xleftrightarrow{*}_{E \cup R} v$ is a sequence

$$\langle \langle u_0, \alpha_1, l_1, r_1, \sigma_1, u_1 \rangle, \langle u_1, \alpha_2, l_2, r_2, \sigma_2, u_2 \rangle, \ldots, \langle u_{n-1}, \alpha_n, l_n, r_n, \sigma_n, u_n \rangle \rangle$$

obtained by concatenating proof steps, with $u = u_0$ and $v = u_n$. It is obvious that proofs can be concatenated or split into two subproofs. A *trivial proof* that $u = v$ is the empty sequence $\langle \rangle$. A proof consisting only of rewrite steps involving rules in R used from left to right is denoted as

$$u_0 \longrightarrow_R u_1 \ldots u_{n-1} \longrightarrow_R u_n$$

or $u_0 \xrightarrow{*}_R u_n$. A proof consisting only of rewrite steps involving rules in R used from right to left is denoted as

$$u_0 \longleftarrow_R u_1 \ldots u_{n-1} \longleftarrow_R u_n$$

or $u_0 \xleftarrow{*}_R u_n$. A proof of the form $u \xrightarrow{*}_R w \xleftarrow{*}_R v$ is called a *rewrite proof*. A proof of the form $u \longleftarrow_R w \longrightarrow_R v$ is called a *peak*. Clearly, a proof is a rewrite proof iff it is a proof without peaks.

We also need the concepts of orientable instance, ground Church-Rosser, and critical pair.

Definition 6.1.2 Let E be a set of equations and $\succ$ a reduction ordering. Given a variant $l \doteq r$ of an equation in $E \cup E^{-1}$, an equation $\sigma(l) \doteq \sigma(r)$ is an *orientable instance* (w.r.t. $\succ$) of $l \doteq r$ iff $\sigma(l) \succ \sigma(r)$ for some substitution σ.[2] Given a reduction ordering $\succ$, the set of all orientable instances of equations in $E \cup E^{-1}$ is denoted by $R(E)$. Note that if $u \longrightarrow_{R(E)} v$, then $u \longrightarrow_{[\alpha, \sigma(l) \doteq \sigma(r)]} v$ for some variant of an equation $l \doteq r$ in $E \cup E^{-1}$ such that $\sigma(l) \succ \sigma(r)$, and since $\succ$ is a reduction ordering, $u \succ v$.

Definition 6.1.3 Let E be a set of equations, R a rewrite system, and $\succ$ a reduction ordering. The pair (E, R) is *ground Church-Rosser relative to* $\succ$ iff (a) $R \subseteq \succ$ and (b) for any two *ground* terms u, v, if $u \stackrel{*}{\longleftrightarrow}_{E \cup R} v$, then there is a rewrite proof

$$u \stackrel{*}{\longrightarrow}_{R(E) \cup R} w \stackrel{*}{\longleftarrow}_{R(E) \cup R} v$$

for some w. A reduction ordering $\succ$ is *total on E-equivalent ground terms* iff for any two distinct ground terms u, v, if $u \stackrel{*}{\longleftrightarrow}_{E} v$, then either $u \succ v$ or $v \succ u$. A reduction ordering $\succ$ that is total on E-equivalent ground terms is called a *ground reduction ordering for E*.

We shall assume in what follows that any reduction ordering is also a simplification ordering; for our purposes what is important is that there exist reduction/simplification orderings, such as the lexicographic path ordering, which are total on ground terms.

It is important to note that for every set R of rewrite rules which is noetherian with respect to a given reduction ordering $\succ$, if R is Church-Rosser, then it is ground Church-Rosser relative to $\succ$, but in general the converse is not true. For example, consider the set of rewrite rules

$$R = \{\ fx \rightarrow gx$$
$$fx \rightarrow hx$$
$$fa \rightarrow a$$
$$ga \rightarrow a$$
$$ha \rightarrow a\ \},$$

where $\Sigma = \{f, g, h, a\}$. It is easy to show that R is noetherian with respect to the recursive path ordering generated by the precedence $f \succ g \succ h \succ a$, and, since every ground term reduces to a, it is ground Church-Rosser

---

[2] As remarked in Chapter 2, because $\succ$ is stable and has the subterm property, for any two terms u and v, $u \succ v$ implies that $Var(v) \subseteq Var(u)$. In the present case thus $Var(\sigma(r)) \subseteq Var(\sigma(l))$.

relative to $\succ$. But R is *not* Church-Rosser, since $hy \longleftarrow_R fy \longrightarrow_R gy$, and hy and gy are irreducible. In general, being Church-Rosser is a stronger condition than being ground Church-Rosser.

Using Lemma 5.3.1, it is easy to show that for any two ground terms u, v, if $u \overset{*}{\longleftrightarrow}_{E \cup R} v$, then there is also a proof Π with sequence of terms $u_0, \ldots, u_n$ where all the u_i are ground. If $\succ$ is a ground reduction ordering for E, then each equality step $u_{i-1} \longleftrightarrow_E u_i$ in the proof Π must be either of the form $u_{i-1} \longrightarrow_{R(E)} u_i$ or $u_{i-1} \longleftarrow_{R(E)} u_i$.

The next definition and lemma extend those given in Section §3.5.2 to this more general context.

Definition 6.1.4 Let E be a set of equations, R a rewrite system, and $\succ$ a reduction ordering containing R. Let $l_1 \rightarrow r_1$ and $l_2 \rightarrow r_2$ be variants of rules in $E \cup E^{-1} \cup R$ with no variables in common (viewing an equation $l \doteq r \in E \cup E^{-1}$ as the rule $l \rightarrow r$). Suppose that for some address β in l_1, l_1/β is *not* a variable and l_1/β and l_2 are unifiable, and let σ be the mgu of l_1/β and l_2. If $\sigma(r_1) \not\succeq \sigma(l_1)$ and $\sigma(r_2) \not\succeq \sigma(l_2)$, the *superposition* of $l_1 \rightarrow r_1$ on $l_2 \rightarrow r_2$ at β determines a *critical pair* g, d of (E, R), with $g = \sigma(r_1)$ and $d = \sigma(l_1[\beta \leftarrow r_2])$. The term $\sigma(l_1)$ is called the *overlapped term*, and β the *critical pair position*.

The importance of critical pairs lies in the fact that they can be used to eliminate peaks in proofs.

Lemma 6.1.5 (Knuth-Bendix, Huet) Let E be a set of equations, R a rewrite system, and $\succ$ a reduction ordering containing R. For every peak $s \longleftarrow_{R(E) \cup R} u \longrightarrow_{R(E) \cup R} t$, either there exists some term v such that $s \overset{*}{\longrightarrow}_{R(E) \cup R} v \overset{*}{\longleftarrow}_{R(E) \cup R} t$, or there exists a critical pair g, d of $E \cup R$, an address α in u (s.t. u/α is not a variable) and a substitution η such that, $s = u[\alpha \leftarrow \eta(g)]$ and $t = u[\alpha \leftarrow \eta(d)]$.

We shall now prove that given a pair (E, R) and a reduction ordering $\succcurlyeq$ containing R that is a ground reduction ordering for $E \cup R$, there is a pair (E^ω, R^ω) containing (E, R) that is equivalent to (E, R) and is ground Church-Rosser relative to $\succcurlyeq$. The pair (E^ω, R^ω) can be viewed as an abstract completion of (E, R) (not produced by any specific algorithm). The existence of (E^ω, R^ω) follows from the existence of fair unfailing completion procedures proved by Bachmair, Dershowitz, Hsiang, and Plaisted [8, 12, 9]. However, this proof requires more machinery than we need for our purposes. We give a more direct and simpler proof (inspired by their proof) that isolates clearly the role played by critical pairs. (In this proof, one will not be distracted by features of completion procedures that have to do with efficiency, like simplification of equations or rules by other rules.) The following definition is needed.

Definition 6.1.6 Let E be a set of equations, R a rewrite system, and $\succ$ a reduction ordering containing R. Let $CR(E, R)$ denote the set of all critical pairs of (E, R) (w.r.t. $\succ$). The sets E^n and R^n are defined inductively as follows: $E^0 = E$, $R^0 = R$, and for every $n \geq 0$,

$$R^{n+1} = R^n \cup \{g \rightarrow d \mid g, d \in CR(E^n, R^n) \text{ and } g \succ d\}$$
$$\cup \{d \rightarrow g \mid g, d \in CR(E^n, R^n) \text{ and } d \succ g\},$$

and

$$E^{n+1} = E^n \cup \{g \doteq d \mid g, d \in CR(E^n, R^n), \ g \not\succeq d \text{ and } d \not\succeq g\}.$$

We also let

$$E^\omega = \bigcup_{n \geq 0} E^n \quad \text{and} \quad R^\omega = \bigcup_{n \geq 0} R^n.$$

Thus, R^ω consists of orientable critical pairs obtained from (E, R) (hereditarily), and E^ω consists of nonorientable critical pairs obtained from (E, R) (hereditarily).

As the next theorem shows, (E^ω, R^ω) is a kind of abstract completion of (E, R).

Theorem 6.1.7 Let E be a set of equations, R a rewrite system, and $\succ$ a reduction ordering containing R that can be extended to a ground reduction ordering $\twoheadrightarrow$ for $E \cup R$. Then, (E^ω, R^ω) is equivalent to (E, R) and is ground Church-Rosser relative to $\twoheadrightarrow$.

Proof. That (E^ω, R^ω) is equivalent to (E, R) follows easily from the fact that (E^ω, R^ω) contains (E, R) and that critical pairs in $CR(E^n, R^n)$ are provably equal from (E^n, R^n). We need to prove that for any two *ground* terms u, v, if $u \xleftrightarrow{\ *\ }_{E^\omega \cup R^\omega} v$, then there is a rewrite proof

$$u \xrightarrow{\ *\ }_{R(E^\omega) \cup R^\omega} w \xleftarrow{\ *\ }_{R(E^\omega) \cup R^\omega} v$$

for some w. Let $\Pi =$

$$\langle\langle u_0, \alpha_1, l_1, r_1, \sigma_1, u_1\rangle, \langle u_1, \alpha_2, l_2, r_2, \sigma_2, u_2\rangle, \ldots, \langle u_{n-1}, \alpha_n, l_n, r_n, \sigma_n, u_n\rangle\rangle$$

be a proof that $u \xleftrightarrow{\ *\ }_{E^\omega \cup R^\omega} v$ (where n is minimal), with $u = u_0$, $v = u_n$, and where u and v are ground. Because $\twoheadrightarrow$ is a ground reduction ordering for $E \cup R$, as observed earlier, we can always assume that the terms u_i are all ground, and we have in fact a proof $u \stackrel{*}{\longleftrightarrow}_{R(E^\omega)\cup R^\omega} v$. We show that for every proof Π of the form $u \stackrel{*}{\longleftrightarrow}_{R(E^\omega)\cup R^\omega} v$, there is a rewrite proof $u \stackrel{*}{\longrightarrow}_{R(E^\omega)\cup R^\omega} w \stackrel{*}{\longleftarrow}_{R(E^\omega)\cup R^\omega} v$, by induction on the multiset $\{u_0, \ldots, u_n\}$, using the multiset ordering $\gg_m$. For the base case, if the rewrite sequence is either trivial (i.e. $u = v$, corresponding to the multiset $\{u\}$) or consists of a single step (corresponding to the multiset $\{u, v\}$), then clearly the proof has no peaks and so is a rewrite proof. For the induction step, suppose Π is a proof with corresponding multiset $\{u_0, \ldots, u_n\}$ with $n \geq 2$. If Π has no peaks, then it is a rewrite proof and we are done. Otherwise, let $u_{i-1} \longleftarrow_{R(E^\omega)\cup R^\omega} u_i \longrightarrow_{R(E^\omega)\cup R^\omega} u_{i+1}$ be a peak in Π. Note that $u_i \gg u_{i-1}$ and $u_i \gg u_{i+1}$ since $R(E^\omega)$ is the set of orientable instances w.r.t. $\gg$ of $E^\omega \cup (E^\omega)^{-1}$, and since R^ω is contained in $\succ$ by its definition. By the critical pair Lemma 6.1.5, either there is some term v such that $u_{i-1} \stackrel{*}{\longrightarrow}_{R(E^\omega)\cup R^\omega} v \stackrel{*}{\longleftarrow}_{R(E^\omega)\cup R^\omega} u_{i+1}$, or $u_{i-1} \longleftrightarrow_{[\eta(g)\doteq\eta(d)]} u_{i+1}$, where $\eta(g) \doteq \eta(d)$ is a ground instance of a critical pair g, d of $E^\omega \cup R^\omega$. In the first case, we can replace the peak by a rewrite proof relative to $\gg$ and we obtain a proof Π' with associated sequence $u_0, \ldots, u_{i-1}, v_1, \ldots, v_k, u_{i+1}, \ldots, u_n$ such that $u_i \gg v_j$ for all j, $1 \leq j \leq k$. Hence

$$\{u_0, \ldots, u_n\} \gg_m \{u_0, \ldots, u_{i-1}, v_1, \ldots, v_k, u_{i+1}, \ldots, u_n\},$$

and we conclude by applying the induction hypothesis. In the second case, observe that $E^\omega \cup R^\omega$ is closed under the formation of critical pairs, and so, $g \doteq d \in E^\omega \cup R^\omega$. Thus, $\eta(g) \doteq \eta(d)$ is orientable either because $g \doteq d \in R^\omega$, or because $g \doteq d \in E^\omega$ and $\gg$ is a ground reduction ordering relative to $E \cup R$. Hence, the peak can be replaced by a proof step $u_{i-1} \longleftrightarrow_{R(E^\omega)\cup R^\omega} u_{i+1}$, obtaining a proof Π' with associated sequence $u_0, \ldots, u_{i-1}, u_{i+1}, \ldots, u_n$. Since

$$\{u_0, \ldots, u_n\} \gg_m \{u_0, \ldots, u_{i-1}, u_{i+1}, \ldots, u_n\},$$

we conclude by applying the induction hypothesis. This concludes the proof. $\square$

Note that since a proof is finite, for any proof $u \stackrel{*}{\longleftrightarrow}_{E^\omega \cup R^\omega} v$, there is a rewrite proof $u \stackrel{*}{\longrightarrow}_{R(E^k)\cup R^k} w \stackrel{*}{\longleftarrow}_{R(E^k)\cup R^k} v$ for some natural number k. Thus, only finitely many critical pairs need to be computed. In some sense, the number of critical pairs to be computed shows how "nonground Church-Rosser" (E, R) is. Also, a sufficient condition for Theorem 6.1.7 to apply is that the reduction ordering $\succ$ containing R be also a total reduction ordering on ground terms. In particular, the theorem applies when $R = \emptyset$, in which case only a total simplification ordering on ground terms is needed. As mentioned earlier, such orderings always exist. On the other hand, given a set R of rewrite rules, there may not be any simplification ordering containing R that is also total on ground terms. Such behavior is illustrated by the set $R = \{f(a) \rightarrow f(b),\ g(b) \rightarrow g(a)\}$.

The fact that a system (E, R) is ground Church-Rosser has the important consequence that $R(E) \cup R$ is canonical on ground terms. This is shown as follows. First, note that $R(E) \cup R$ is Noetherian on ground terms, since R is contained in the reduction ordering $\succ$ by hypothesis and $R(E)$ is also contained in $\succ$ since it is the set of orientable instances of E relative to $\succ$ (which is total on ground terms). To show confluence, note that for any *ground* terms u, v_1, v_2, if

$$v_1 \xleftarrow{\ *\ }_{R(E)\cup R} u \xrightarrow{\ *\ }_{R(E)\cup R} v_2,$$

then $v_1 \xleftrightarrow{\ *\ }_{R(E)\cup R} v_2$, and since (E, R) is ground Church-Rosser, there is a rewrite proof

$$v_1 \xrightarrow{\ *\ }_{R(E)\cup R} w \xleftarrow{\ *\ }_{R(E)\cup R} v_2$$

for some w. Hence, every ground term u can be reduced to a unique normal form $u\!\downarrow$ (w.r.t. $R(E) \cup R$).

It is very useful to observe that if a procedure P for finding sets of E-unifiers satisfies the property stated in the next definition, then in order to show that this procedure yields complete sets, there is no loss of generality in showing completeness with respect to *ground* E-unifiers whose domains contain $Var(S)$ (that is, in clause (iii) of Definition 4.1.6, $\theta(x)$ is a ground term for every $x \in D(\theta)$, and $Var(S) \subseteq D(\theta)$).

Definition 6.1.8 We call an E-unification procedure P *pure* if for every ranked alphabet Σ, every finite set E of equations over $T_\Sigma(X)$ and every equation system S over $T_\Sigma(X)$, if $U = P(E, S)$ is the set of E-unifiers for S given by procedure P, then for every $\sigma \in U$, for every $x \in D(\sigma)$, every constant or function symbol occurring in $\sigma(x)$ occurs either in some equation in E or some equation in S.

In other words, $P(E, S)$ does not contain constant or function symbols that do not already occur in the input (E, S). (For example, it is easy to prove that all the sets of transformations presented in this monograph are pure.) To prove our previous claim, we proceed as follows. We add countably infinitely many new (distinct) constants c_x to Σ, each constant c_x being associated with the variable x. The resulting alphabet is denoted by Σ_{SK}. If θ is not ground, we create the Skolemized version of θ, that is, the substitution $\widehat{\theta}$ obtained by replacing the variables in the terms $\theta(x)$ by new (distinct) constants.[3]

Lemma 6.1.9 Given a pure E-unification procedure P, assume that for every ranked alphabet Σ, every finite set E of equations over $T_\Sigma(X)$ and every S over $T_\Sigma(X)$, the set $U = P(E, S)$ of E-unifiers of S given by P satisfies conditions (i) and (ii) of Definition 4.1.6, and for every E-unifier θ of S such that $Var(S) \subseteq D(\theta)$ and $\theta(x) \in T_\Sigma$ for every $x \in D(\theta)$, there is some $\sigma \in U$ such that $\sigma \leq_E \theta[Var(S)]$ (c.f. condition (iii) of Definition 4.1.6). Then every set $U = P(E, S)$ is a complete set of E-unifiers for S.

Proof. Let θ be any E-unifier of S over $T_\Sigma(X)$. If $D(\theta)$ does not contain $Var(S)$, extend θ such that $\theta(y) = c_y$ for every $y \in Var(S) - D(\theta)$, and let $\widehat{\theta}$ be the Skolemized version of this extension of θ. We are now considering the extended alphabet Σ_{SK}. It is immediately verified that $\widehat{\theta}$ is also an E-unifier of S such that $Var(S) \subseteq D(\widehat{\theta})$ and $\widehat{\theta}(x) \in T_{\Sigma_{SK}}$ for all $x \in D(\widehat{\theta})$. Then, there is some $\sigma \in U$ such that $\sigma \leq_E \widehat{\theta}[Var(S)]$, which means that there is some substitution η (over $T_{\Sigma_{SK}}(X)$) such that $\sigma \circ \eta =_E \widehat{\theta}[Var(S)]$. Note that by the purity of P, since E and S do not contain Skolem constants, σ does not contain Skolem constants. Let η' be obtained from η by changing each Skolem constant back to the corresponding variable. Since σ does not contain Skolem constants, it is immediately verified that $\sigma \circ \eta' =_E \theta$. Thus, the set U is a complete set of E-unifiers for S over $T_\Sigma(X)$. $\square$

The following result is also useful.

Lemma 6.1.10 Let E be a set of equations, R a rewrite system, and $\succ$ a reduction ordering containing R, and assume that (E, R) is ground Church-Rosser relative to $\succ$. If θ is a ground (E, R)-unifier of u and v and $Var(u, v) \subseteq D(\theta)$, then there is a ground substitution σ that is reduced w.r.t. $R(E) \cup R$ such that $\sigma =_{EUR} \theta$, σ is an (E, R)-unifier of u and v, and $Var(u, v) \subseteq D(\sigma)$.

Proof. Since (E, R) is ground Church-Rosser relative to $\succ$, $R(E) \cup R$ is canonical on ground terms. Thus, if $\theta(u) \overset{*}{\longleftrightarrow}_{EUR} \theta(v)$, since θ is ground and $Var(u, v) \subseteq D(\theta)$, then there is a rewrite proof

$$\theta(u) \overset{*}{\longrightarrow}_{R(E)\cup R} u' \overset{*}{\longrightarrow}_{R(E)\cup R} w \overset{*}{\longleftarrow}_{R(E)\cup R} v' \overset{*}{\longleftarrow}_{R(E)\cup R} \theta(v)$$

where w is ground and in normal form (w.r.t. $R(E) \cup R$), and where the reductions $\theta(u) \overset{*}{\longrightarrow}_{R(E)\cup R} u'$ and $v' \overset{*}{\longleftarrow}_{R(E)\cup R} \theta(v)$ reduce each $\theta(x)$

---

[3] More precisely, $\widehat{\theta}$ is obtained from θ by replacing every variable y in each term $\theta(x)$ by the corresponding Skolem constant c_y, for each $x \in D(\theta)$.

$(x \in D(\theta))$ to its normal form $\theta(x){\downarrow}$ (w.r.t. $R(E) \cup R$). Thus, defining the reduced substitution σ such that $\sigma(x) = \theta(x){\downarrow}$ for each $x \in D(\theta)$, we have $u' = \sigma(u)$, $v' = \sigma(v)$, σ is a ground (E, R)-unifier of u and v, and $\sigma =_{E \cup R} \theta$. $\square$

For our next result, we need the following definition.

Definition 6.1.11   Given a rewrite system R, a step $u \longrightarrow_{[\beta, l \doteq r, \rho]} v$ is *innermost* (w.r.t. R) iff every proper subterm of $u/\beta = \rho(l)$ is irreducible w.r.t. R.

The next lemma shows that in ground Church-Rosser systems, normal forms can always be reached via certain canonical innermost rewrite sequences. The proof is not trivial because $Var(r) - Var(l)$ may be nonempty for an equation $l \doteq r \in E \cup E^{-1}$.

Lemma 6.1.12   Let E be a set of equations, R a rewrite system, and $\succ$ a reduction ordering containing R, and assume that (E, R) is ground Church-Rosser relative to $\succ$. Every ground term u reduces to its normal form $u{\downarrow}$ (w.r.t. $R(E) \cup R$) in a sequence of innermost reductions $u \xrightarrow{\;*\;}_{R(E) \cup R} u{\downarrow}$, such that for every rule $\rho(l) \to \rho(r)$ used in the sequence, ρ is reduced (w.r.t. $R(E) \cup R$).

Proof. Since (E, R) is ground Church-Rosser relative to $\succ$, $R(E) \cup R$ is canonical on ground terms. We proceed by induction on the well founded ordering $\succ$. If u is in normal form, we are done. Otherwise, there is a sequence of reduction steps $u \xrightarrow{\;*\;}_{R(E) \cup R} u{\downarrow}$, and and because u is ground, we can assume that every rule $\rho(l) \to \rho(r)$ used in such a proof is ground. Note that $\rho(l) \succ \rho(r)$ whenever either $l \to r \in R$ or $\rho(l) \to \rho(r) \in R(E)$, and $Var(l) \cup Var(r) = D(\rho)$ since $\rho(l)$ and $\rho(r)$ are ground.[4] If u is not in normal form, there must be some innermost step

$$u \longrightarrow_{[\beta, l \doteq r, \rho]} u[\beta \leftarrow \rho(r)].$$

For every $x \in Var(l)$, $\rho(x)$ must be in normal form (w.r.t. $R(E) \cup R$), since otherwise some proper subterm of $\rho(l) = u/\beta$ would be reducible, contradicting the fact that we have an innermost step. For each $x \in (Var(r) - Var(l))$, let $\rho(x){\downarrow}$ be the normal form of $\rho(x)$ (w.r.t. $R(E) \cup R$), and let ρ' be the reduced substitution such that $\rho'(x) = \rho(x){\downarrow}$ for each

---

[4] Certainly, $\rho(l)$ and $\rho(r)$ ground implies that $Var(l) \cup Var(r) \subseteq D(\rho)$, but the fact that ρ may be defined outside of $Var(l) \cup Var(r)$ is not used anywhere, so we might as well assume that $Var(l) \cup Var(r) = D(\rho)$.

$x \in (Var(r) - Var(l))$, and $\rho'(x) = \rho(x)$ for each $x \in Var(l)$. The definition of ρ' implies that $\rho'(l) = \rho(l)$ and $\rho(x) \succeq \rho'(x)$ for every $x \in D(\rho)$. Thus, $\rho(l) \succ \rho(r)$ implies that $\rho'(l) \succ \rho'(r)$. Since $R(E) \cup R$ is canonical on ground terms, $\rho'(l) = \rho(l)$, and $u = u[\beta \leftarrow \rho(l)]$, using the rule $\rho'(l) \to \rho'(r)$, we have a proof

$$u = u[\beta \leftarrow \rho'(l)] \longrightarrow_{R(E) \cup R} u[\beta \leftarrow \rho'(r)] \overset{*}{\longrightarrow}_{R(E) \cup R} u{\downarrow}$$

where the first reduction step is innermost and ρ' is reduced (w.r.t. $R(E) \cup R$). Letting $u' = u[\beta \leftarrow \rho'(r)]$, we have $u \succ u'$ since $\rho'(l) \succ \rho'(r)$. We conclude by applying the induction hypothesis to u'. $\quad\square$

This leads naturally to the following definition, which will be fundamental to our completeness proof.

Definition 6.1.13 Let (E, R) and $\succ$ be as in the preceeding lemma. If θ is an (E, R)-irreducible substitution which is a ground (E, R)-unifier of two terms s and t, then a *basic rewrite proof* of this fact is a rewrite proof

$$\theta(s) \overset{*}{\longrightarrow}_{R(E) \cup R} w \overset{*}{\longleftarrow}_{R(E) \cup R} \theta(t)$$

for some w, where for each instance $\rho(l) \to \rho(r)$ used in the proof, ρ is reduced w.r.t. $R(E) \cup R$.

Note that we do not require w to be irreducible, since perhaps there is a proof of minimal length in which this is not the case. But the lemma implies that such a basic rewrite proof always exists. The essential fact is that no reduction takes place at any substitution position in the sequence.

6.2 Completeness of the Set $\mathcal{T}$

Let E be a set of equations, R a rewrite system, and $\succ$ a reduction ordering containing R.

Definition 6.2.1 (The set of transformation rules $\mathcal{T}$) The set $\mathcal{T}$ consists of the transformations Trivial, Term Decomposition, and Variable Elimination from the set $\mathcal{ST}$ plus one more transformation defined as follows:

Lazy Paramodulation: Given a multiset of equations $\{u \approx v\} \cup S$, then

$$\{u \approx v\} \cup S \Longrightarrow \{u/\beta \approx l, u[\beta \leftarrow r] \approx v\} \cup S, \qquad (4a)$$

where β is a nonvariable occurrence in u (i.e., u/β is *not* a variable) and $l \doteq r$ is a variant (whose variables do not occur in $\{u \approx v\} \cup S$) of some equation in $E \cup E^{-1} \cup R \cup R^{-1}$. Furthermore, if l is not a variable, then $Root(u/\beta) = Root(l)$ and Term Decomposition is immediately applied to $u/\beta \approx l$ (this corresponds to a leftmost rewrite at address β).[5] Thus, if l is not a variable, letting $l = f(l_1, \ldots, l_k)$ and $u/\beta = f(t_1, \ldots, t_k)$, Lazy Paramodulation can be specialized to:

$$\{u \approx v\} \cup S \Longrightarrow \{t_1 \approx l_1, \ldots, t_k \approx l_k, u[\beta \leftarrow r] \approx v\} \cup S. \qquad (4a')$$

Recall that an equation $u \approx v$ is in fact a multiset, and so Lazy Paramodulation also applies from v to u, as in

$$\{u \approx v\} \cup S \Longrightarrow \{v/\beta \approx l, u \approx v[\beta \leftarrow r]\} \cup S, \qquad (4a)$$

where β is a nonvariable occurrence in v. As in our previous set of transformations, we note that systems are multisets and the unions in this rule are multiset unions.

In order to distinguish between the set $\mathcal{BT}$ and the set $\mathcal{T}$, the former will be called $\mathcal{BT}$-transformations and the latter $\mathcal{T}$-transformations. The soundness of the $\mathcal{T}$-transformations is given by

Theorem 6.2.2 (Soundness of $\mathcal{T}$) If $S \overset{*}{\Longrightarrow} S'$, using transformations from the set $\mathcal{T}$, with S' in solved form, then $\sigma_{S'}|_{Var(S)} \in U_E(S)$.

Proof. The only difference from Theorem 5.2.3 is that we must prove the soundness of Lazy Paramodulation, i.e., that if $S \Longrightarrow S'$ using this transformation, then $U_E(S') \subseteq U_E(S)$. But clearly if $\theta(u/\beta) \overset{*}{\longleftrightarrow}_E \theta(l)$ and $\theta(u[\beta \leftarrow r]) \overset{*}{\longleftrightarrow}_E \theta(v)$ then we have

$$\begin{aligned}
\theta(u) &= \theta(u[\beta \leftarrow u/\beta]) \\
&\overset{*}{\longleftrightarrow}_E \theta(u[\beta \leftarrow l]) \\
&\underset{[\beta, l \doteq r, \theta]}{\longrightarrow} \theta(u[\beta \leftarrow r]) \\
&\overset{*}{\longleftrightarrow}_E \theta(v),
\end{aligned}$$

from which the result follows.   □

The completeness of the set of $\mathcal{T}$-transformations is shown in two steps. First, we assume that (E, R) is ground Church-Rosser and we show that the $\mathcal{T}$-transformations are complete, even when Lazy Paramodulation is

---

[5] As with Root Rewriting, note that this is *not* simply a paramodulation step, nor simply a paramodulation step where the unification of u/β and l is delayed; it allows further rewrite steps to occur below (but *not* at) the roots of u/β and l, hence the name *Lazy* Paramodulation. In Section §6.5 this restriction will be strengthened.

restricted so that it applies only when either $\beta = \epsilon$ (i.e. at the root) or when one of u, v is a variable (but not both). Then, we use Theorem 6.1.7 and a lemma that shows that the computation of critical pairs can be simulated by Lazy Paramodulation unrestricted.

The first part consists in showing that we can define a certain kind of presentation that a substitution θ is an (E, R)-unifier of a system S. By induction on the complexity of this representation, we can extract the appropriate sequence of transformations.

Definition 6.2.3 Let E be a set of equations, R a rewrite system, and $\succ$ a reduction ordering containing R, such that (E, R) is ground Church-Rosser relative to $\succ$. Furthermore, let θ be an (E, R)-irreducible substitution which is an (E, R)-unifier of a system S, and such that $Var(S) \subseteq D(\theta)$. A *basic proof system* for θ, (E, R), and S is a triple $\langle \theta, S, \Gamma \rangle$, where Γ is a multiset of basic proofs such that there exists a basic rewrite proof of minimal length $\theta(s) \xrightarrow{*}_{R(E) \cup R} w \xleftarrow{*}_{R(E) \cup R} \theta(t)$ in Γ corresponding to every $s \approx t \in S$.

The complexity $\mu(\langle \theta, S, \Gamma \rangle)$ of a basic proof system is a quadruple $\langle n_1, n_2, n_3 \rangle$ where

(i) n_1 is the number of reduction steps occurring in all the equational proofs in Γ;

(ii) n_2 is the number of unsolved variables in S; and

(iii) n_3 is the sum of the sizes of all terms occurring in S.

The ordering associated with this complexity measure is the lexicographic ordering using the standard ordering on the natural numbers for each component. (Cf. the ordering used for proving the completeness of the set ST.)

The fact that such proof systems always exist is easily proved, given the results above.

Proposition 6.2.4 If (E, R) is ground Church-Rosser relative to $\succ$ and θ is a ground, irreducible (E, R)-unifier of a system S, where $Var(S) \subseteq D(\theta)$, then there exists a basic proof system for θ, (E, R), and S.

We are now in the position to prove the completeness of the set T when (E, R) is ground Church-Rosser.

Lemma 6.2.5 Let E be a set of equations, R a rewrite system, and $\succ$ a reduction ordering containing R, and assume that (E, R) is ground Church-Rosser relative to $\succ$. Given any system S if θ is an (E, R)-unifier of S, then there is a sequence of T-transformations $S \overset{*}{\Longrightarrow} T$ (using variants of equations in $E \cup E^{-1} \cup R$) yielding a solved system T such that if σ_T is the substitution associated with T, then $\sigma_T \leq_{EUR} \theta[Var(S)]$. Furthermore, Lazy Paramodulation can be restricted so that it is applied only when either $\beta = \epsilon$ or one of u, v is a variable (but not both).

Proof. First, observe that any procedure using the transformations in T satisfies the purity condition of Definition 6.1.8, and by Lemma 6.1.10 and Lemma 6.1.9, we can assume that θ is reduced w.r.t. $R(E) \cup R$, ground, and that $Var(S) \subseteq D(\theta)$. Thus there must exist a basic proof system $\Pi = \langle \theta, S, \Gamma \rangle$; we construct a requisite sequence of transformations by induction on the complexity of this proof system.

The base case occurs when S is in solved form: the trivial transformation sequence of length 0, viz. $S = T$, yields a substitution σ_T, and by Lemma 5.1.1 we have that $\sigma_T \leq_{EUR} \theta[Var(S)]$.

For the induction step, suppose that S is not in solved form, so that if $\mu(\Pi) = \langle n_1, n_2, n_3 \rangle$, then $n_2 \neq 0$. Then there must exist some $s \approx t \in S$ which is not in solved form. There are several cases (not all of which are mutually exclusive).

(A) If in fact $s = t$, then we can apply Trivial to obtain a new system S' which has a basic proof system $\langle \theta, S', \Gamma - \{ \langle \rangle \} \rangle$ of strictly smaller complexity than Π, since n_1 and n_2 can not increase and n_3 must decrease. By the induction hypothesis, there exists a proof $S \Longrightarrow_{triv} S' \overset{*}{\Longrightarrow} T$ such that $\sigma_T \leq_{EUR} \theta[Var(S')]$, i.e., there exists some η such that $\sigma_T \eta =_{EUR} \theta[Var(S')]$. Now perhaps $V = Var(S) - Var(S')$ is non-empty. But if so, since all variables introduced in the proof $S' \overset{*}{\Longrightarrow} T$ are new, $V \cap Var(T) = \emptyset$, so $V \cap (D(\sigma_T) \cup I(\sigma_T)) = \emptyset$, but $V \subseteq D(\theta)$. Since we can assume wlg that $D(\eta) \cap V = \emptyset$, then $\theta =_{EUR} \sigma_T \eta \theta[Var(S)]$, since for any $x \in Var(S')$, $\theta(x) =_{EUR} \eta(\sigma_T(x)) = \theta(\eta(\sigma_T(x)))$ because $\eta(\sigma_T(x))$ is ground, and for any $y \in V$, $\eta(\sigma_T(y)) = y$, so that $\theta(y) = \theta(\eta(\sigma_T(y)))$. Thus $\sigma_T \leq_{EUR} \theta[Var(S)]$.

(B) If $s \approx t$ is in the form $f(s_1, \ldots, s_k) \approx f(t_1, \ldots, t_k)$ for some $f \in \Sigma_k$, and $\theta(s) = \theta(t)$, then we can apply Decomposition, and clearly we have a new computed answer proof $\langle \theta, S', \Gamma' \rangle$ where we add $k - 1$ trivial equational proofs $\langle \rangle$ to Γ to obtain Γ'; the complexity has decreased because n_3 has decreased, but n_2 and n_1 can not increase; since $Var(S) = Var(S')$, we conclude as in the previous case.

(C) If $s \approx t$ is in the form $x \approx t$ where $\theta(x) = \theta(t)$, then we can apply Variable Elimination, and if $S = \{x \approx t\} \cup S''$ and $\sigma = [t/x]$, then $\theta = \sigma\theta$ (since $\theta(x) = \theta(t) = \theta(\sigma(x))$ and $\theta(y) = \theta(\sigma(y))$ for $y \neq x$). Therefore $\theta(\{x \approx t\} \cup S'') = \theta(\{x \approx t\} \cup \sigma(S''))$. Thus there exists a new proof $\langle \theta, S', \Gamma \rangle$ with a strictly smaller complexity, since n_1 is unchanged but n_2 has decreased. Since $Var(S) = Var(S')$, we conclude as above.

(D) Finally, in the case that $\theta(s) \neq \theta(t)$, if we assume wlg that $\theta(s) \succ \theta(t)$, then there must exist a non-trivial basic rewrite proof

$$\theta(s) \xrightarrow{*}_{R(E)\cup E} \xleftarrow{*}_{R(E)\cup E} \theta(t)$$

in Γ. If there is no rewrite at the root, then suppose $s = f(s_1, \ldots, s_k)$ and $t = f(t_1, \ldots, t_k)$ for some $f \in \Sigma_k$. We may apply Decomposition to obtain a new system S' and also we may produce a new set Γ' of basic proofs from Γ by replacing the proof $\theta(s) \xrightarrow{*}_{R(E)\cup E} \xleftarrow{*}_{R(E)\cup E} \theta(t)$ by the k proofs obtained by decomposition, i.e., $\theta(s_i) \xrightarrow{*}_{R(E)\cup E} \xleftarrow{*}_{R(E)\cup E} \theta(t_i)$ for $1 \leq i \leq k$ (the fact that rewrites at disjoint addresses can always be commuted allows this decomposition). It is not to hard to see that these new proofs are also basic, since the matching substitutions used at each step are still reduced. The complexity of the resultant $\langle \theta, S', \Gamma' \rangle$ is strictly smaller than Π, since n_1 and n_2 have not changed, but n_3 has decreased. Since $Var(S) = Var(S')$, we are done.

If there is a rewrite at the root, then wlg assume it occurs in the rewrite sequence from $\theta(s)$, the other case being similar. Then our sequence has the form

$$\theta(s) \xrightarrow{*}_{R(E)\cup E} \rho(l) \xrightarrow{}_{[\epsilon, \rho(l \doteq r)]} \rho(r) \xrightarrow{*}_{R(E)\cup E} \xleftarrow{*}_{R(E)\cup E} \theta(t)$$

which, by the assumption that all equations and rewrite rules used are variants whose variables are disjoint from the rest of the system, can be equivalently expressed in the form

$$\theta'(s) \xrightarrow{*}_{R(E)\cup E} \theta'(l) \xrightarrow{}_{[\epsilon, \theta'(l \doteq r)]} \theta'(r) \xrightarrow{*}_{R(E)\cup E} \xleftarrow{*}_{R(E)\cup E} \theta'(t)$$

where $\theta' = \theta\rho$. (For notational convenience, we shall not here distinguish between an equation $l \doteq r \in E$ and a rewrite rule $l \rightarrow r \in R$.) It is not hard to see that the two rewrite sequences

$$\theta'(s) \xrightarrow{*}_{R(E)\cup E} \theta'(l)$$

and

$$\theta'(r) \xrightarrow{*}_{R(E)\cup E} \xleftarrow{*}_{R(E)\cup E} \theta'(t)$$

are still basic, since the matching substitutions are still reduced after breaking apart the former proof.

Thus, we can use Lazy Paramodulation to transform $S = \{s \approx t\} \cup S''$ into $S' = \{s \approx l, r \approx t\} \cup S''$, and, since we can assume $D(\rho) \cap Var(S'') = \emptyset$, there exists a new proof system $\langle \theta', S', \Gamma' \rangle$ for θ', (E, R), and S', where Γ' is produced from Γ by removing the former proof for $s \approx t$ and replacing by the two proofs given above. This is still a basic proof system, since the rewrite proofs added are still basic, and clearly the rest of Γ is essentially unaffected by extending θ to θ'. The complexity of Γ' is strictly less than that of Γ, since we have decreased n_1 by 1. By applying the induction hypothesis, we obtain a proof $S \Longrightarrow_{lp} S' \stackrel{*}{\Longrightarrow} T$ such that $\sigma_T \leq_{E \cup R} \theta'[Var(S')]$. Since $Var(S) \subseteq Var(S')$ and $\theta' = \theta[Var(S)]$, we have $\sigma_T \leq_{E \cup R} \theta[Var(S)]$.

Finally, it should be clear from the previous case that Lazy Paramodulation can be restricted so that it need only be applied below the root when one of the terms in the equation being solved is a non-variable. This is because for an equation $u \approx v$, if v is a variable, say y, because θ is reduced we must have $w = \theta(y)$ and $\theta(u) \stackrel{*}{\longrightarrow}_{R(E) \cup R} \theta(y)$. If u is also a variable, say x, we must have $\theta(x) = w = \theta(y)$. Thus, when $u \approx v$ is an equation consisting of two variables, Trivial or Variable Elimination always applies. $\square$

Note that the case analysis in this lemma is not mutually exclusive (i.e., some equation in S may fit more than one case), but that this does not affect the result.

The next result provides the motivation for our assumption that (E, R) is consistent, and will be used in second part of our completeness result.

Lemma 6.2.6 If (E, R) is consistent and $\langle \sigma(r_1), \sigma(l_1[\beta \leftarrow r_2]) \rangle$ is a critical pair in $CR(E^n, R^n)$ for some n, then both l_1 and l_2 are non-variable terms.

Proof. The term l_1 is a non-variable by definition. If l_2 is a variable, then, since (E, R) is consistent and (E^ω, R^ω) is equivalent to (E, R), (E^ω, R^ω) is also consistent. But then r_2 must contain l_2, and this violates the ordering condition for critical pairs, since we assumed $\succ$ is a simplification ordering. $\square$

In order to prove the completeness of the $\mathcal{T}$-transformations in the general case, the following lemma showing that the computation of critical pairs can be simulated by Lazy Paramodulation is needed.

Lemma 6.2.7 Let E be a set of equations, R a rewrite system, and $\succ$ a reduction ordering containing R. For every finite system S, every sequence of $\mathcal{T}$-transformations $S \stackrel{*}{\Longrightarrow} \widehat{S}$ to a solved system $\widehat{S}$ using equations in $E^\omega \cup (E^\omega)^{-1} \cup R^\omega$ can be converted into a sequence $S \stackrel{*}{\Longrightarrow} \widehat{S'}$ using equations only in $E \cup E^{-1} \cup R \cup R^{-1}$, such that $\widehat{S'}$ is in solved form and $\sigma_{\widehat{S}} = \sigma_{\widehat{S'}}[Var(S)]$.

Proof. Let the *depth* of an equation[6] $l \doteq r$ in $E^{\omega} \cup (E^{\omega})^{-1} \cup R^{\omega}$ be the least k such that $l \doteq r$ is in $E^k \cup (E^k)^{-1} \cup R^k$. We proceed by induction on the (finite) multiset of the depths of all clauses used in the proof $S \overset{*}{\Longrightarrow} \hat{S}$. The base case occurs when the depth of each clause used is 0, and the result holds trivially. Now if the proof uses an equation $\sigma(r_1) \doteq \sigma(l_1[\beta \leftarrow r_2])$ of depth greater than 0, obtained by forming a critical pair from $l_1 \doteq r_1$ and $l_2 \doteq r_2$ at β in l_1, with mgu σ of l_1/β and l_2 (note that each of these component equations must be of smaller depth), then we can show that the original use of the critical pair in a Lazy Paramodulation step can be simulated by two Lazy Paramodulation steps involving the component equations, plus some number of ST transformations to (effectively) compute the critical pair "on the fly." There are several cases, depending on the form of the critical pair and how it was used in the proof.

Case One. Suppose $\sigma(r_1)$ is a variable (which implies that r_1 is also a variable) and we did the Lazy Paramodulation step from $\sigma(r_1)$; then the proof must have the form[7]

$$S \overset{*}{\Longrightarrow} u \approx v, T$$
$$\Longrightarrow_{lp} u/\alpha \approx \sigma(r_1), u[\alpha \leftarrow \sigma(l_1[\beta \leftarrow r_2])] \approx v, T$$
$$\overset{*}{\Longrightarrow} \hat{S}$$

where u/α is a non-variable (note that the decomposition step which forms the second half of the Lazy Paramodulation rule is not applied since $\sigma(r_1)$ is a variable). We can start the simulation of this proof step as follows:

$$S \overset{*}{\Longrightarrow} u \approx v, T$$
$$\Longrightarrow_{lp} u/\alpha \approx r_1, u[\alpha \leftarrow l_1] \approx v, T$$

using the equation $l_1 \doteq r_1$ backwards at α in u. Next, we have

$$u/\alpha \approx r_1, u[\alpha \leftarrow l_1] \approx v, T$$
$$\Longrightarrow_{lp} u/\alpha \approx r_1, Q, u[\alpha \leftarrow l_1[\beta \leftarrow r_2]] \approx v, T$$

using the equation $l_2 \doteq r_2$ at $\alpha\beta$ in $u[\alpha \leftarrow l_1]$ and where Q is either $l_1/\beta \approx l_2$ or the result of a decomposition step, i.e., $l_1/\beta \approx l_2 \Longrightarrow_{dec} Q$. Also, we use the fact that

$$u[\alpha \leftarrow l_1]/\alpha\beta = l_1/\beta$$

---

[6] For notational convenience, we shall not here distinguish between an equation $l \doteq r$ and a rewrite rule $l \rightarrow r$.

[7] Again for notational convenience, we represent here a system in the form $\{u \approx v\} \cup T$ by $u \approx v, T$, etc.

and

$$u[\alpha \leftarrow l_1][\alpha\beta \leftarrow r_2] = u[\alpha \leftarrow l_1[\beta \leftarrow r_2]].$$

Finally, since we assume in this monograph that all *mgus* are such as produced by the set $\mathcal{ST}$, there is a sequence of $\mathcal{ST}$-steps

$$Q \overset{*}{\Longrightarrow}_{st} S_\sigma \qquad\qquad (*)$$

calculating a solved system S_σ representing σ. Now, when we apply these same transformations in the context of the whole sequence, plus (possibly) some additional $\mathcal{ST}$-steps to solve variables which are already solved when they first appear in $(*)$ (and hence do not need to be explicitly eliminated), but which appear in $l[\beta \leftarrow r_2]$, the overall effect is to calculate S_σ and also to apply σ everywhere. But since wlg we may assume that $D(\sigma)$ is disjoint from the set of all variables used elsewhere in the proof, we have for the next part of our new proof sequence:

$$u/\alpha \approx r_1, Q, u[\alpha \leftarrow l_1[\beta \leftarrow r_2]] \approx v, T$$
$$\overset{*}{\Longrightarrow}_{st} u/\alpha \approx \sigma(r_1), S_\sigma, u[\alpha \leftarrow \sigma(l_1[\beta \leftarrow r_2])] \approx v, T.$$

Now if we add S_σ to the proof at this point, since all its equations are solved, and no transformation can *unsolve* a pair, the only thing that will happen is that perhaps some equation $x \approx t \in S_\sigma$ will change into $x \approx t'$ during a later Variable Elimination step. Thus the subsystem S_σ may evolve into some $S_{\sigma'}$, but still we will have $D(\sigma) = D(\sigma')$ and $D(\sigma') \cap Var(S) = \emptyset$. Thus we may finish our new proof with the same remaining sequence of steps as in the original proof:

$$u/\alpha \approx \sigma(r_1), S_\sigma, u[\alpha \leftarrow \sigma(l_1[\beta \leftarrow r_2])] \approx v, T$$
$$\overset{*}{\Longrightarrow} \widehat{S} \cup S_{\sigma'}.$$

where $\sigma_{(S_{\sigma'} \cup \widehat{S})} = \sigma_{\widehat{S}}[Var(S)]$. We conclude by applying the induction hypothesis.

Case Two. Suppose that r_1 is not a variable (which implies that $\sigma(r_1)$ is not a variable) and the Lazy Paramodulation step is again from $\sigma(r_1)$. In this case $Head(u/\alpha) = Head(r_1)$; suppose that $u/\alpha = f(u_1, \ldots, u_n)$ and $r_1 = f(r'_1, \ldots, r'_n)$. Then our proof must have the form

$$S \overset{*}{\Longrightarrow} u \approx v, T$$
$$\Longrightarrow_{lp} u_1 \approx \sigma(r'_1), \ldots, u_n \approx \sigma(r'_n), u[\alpha \leftarrow \sigma(l_1[\beta \leftarrow r_2])] \approx v, T$$
$$\overset{*}{\Longrightarrow} \widehat{S}$$

where u/α is a non-variable. The construction of the new proof proceeds analogously with the previous case, resulting in

$$S \stackrel{*}{\Longrightarrow} u \approx v, T$$
$$\Longrightarrow_{lp} u_1 \approx r_1', \ldots, u_n \approx r_n', u[\alpha \leftarrow l_1] \approx v, T$$
$$\Longrightarrow_{lp} u_1 \approx r_1', \ldots, u_n \approx r_n', Q, u[\alpha \leftarrow l_1[\beta \leftarrow r_2]] \approx v, T$$
$$\stackrel{*}{\Longrightarrow}_{st} u_1 \approx \sigma(r_1'), \ldots, u_n \approx \sigma(r_n'), S_\sigma, u[\alpha \leftarrow \sigma(l_1[\beta \leftarrow r_2])] \approx v, T$$
$$\stackrel{*}{\Longrightarrow} \widehat{S} \cup S_{\sigma'}.$$

Case Three. Suppose the Lazy Paramodulation step still proceeds from $\sigma(r_1)$, but r_1 is a variable and $\sigma(r_1)$ is not a variable. Here the simulation is slightly more delicate. Suppose that $u/\alpha = f(u_1, \ldots, u_n)$ and $\sigma(r_1) = f(v_1, \ldots, v_n)$, so that our original proof was in the form

$$S \stackrel{*}{\Longrightarrow} u \approx v, T$$
$$\Longrightarrow_{lp} u_1 \approx v_1, \ldots, u_n \approx v_n, u[\alpha \leftarrow \sigma(l_1[\beta \leftarrow r_2])] \approx v, T$$
$$\stackrel{*}{\Longrightarrow} \widehat{S}$$

We start our simulation as in case one,

$$S \stackrel{*}{\Longrightarrow} u \approx v, T$$
$$\Longrightarrow_{lp} u/\alpha \approx r_1, u[\alpha \leftarrow l_1] \approx v, T$$
$$\Longrightarrow_{lp} u/\alpha \approx r_1, Q, u[\alpha \leftarrow l_1[\beta \leftarrow r_2]] \approx v, T$$
$$\stackrel{*}{\Longrightarrow}_{st} u/\alpha \approx \sigma(r_1), S_\sigma, u[\alpha \leftarrow \sigma(l_1[\beta \leftarrow r_2])] \approx v, T,$$

but now, since the leftmost pair was in fact decomposed by our original proof, we add an additional decomposition step before proceeding:

$$u/\alpha \approx \sigma(r_1), S_\sigma, u[\alpha \leftarrow \sigma(l_1[\beta \leftarrow r_2])] \approx v, T$$
$$\Longrightarrow_{dec} u_1 \approx v_1, \ldots, u_n \approx v_n, S_\sigma, u[\alpha \leftarrow \sigma(l_1[\beta \leftarrow r_2])] \approx v, T$$
$$\stackrel{*}{\Longrightarrow} \widehat{S} \cup S_{\sigma'}.$$

In other respects this case is analogous to the first two.

This covers all the cases where the Lazy Paramodulation step proceeds from $\sigma(r_1)$.

Case Four. Now suppose that the Lazy Paramodulation step proceeds from $\sigma(l_1[\beta \leftarrow r_2])$. If $\beta = \epsilon$, then by Lemma 6.2.6 we may assume that both l_1 and l_2 are non-variables, and so this possibility is covered by the previous three cases. Thus suppose $\beta = i\beta'$ for $1 \le i \le n$ and let $u/\alpha =$

$f(u_1, \ldots, u_n)$ and $l_1 = f(l'_1, \ldots, l'_n)$. In this case our original proof must be in the form

$$S \stackrel{*}{\Longrightarrow} u \approx v, T$$
$$\Longrightarrow_{lp} u_1 \approx \sigma(l'_1), \ldots, u_i \approx \sigma(l'_i[\beta' \leftarrow r_2]), \ldots, u_n \approx \sigma(l'_n),$$
$$u[\alpha \leftarrow \sigma(r_1)] \approx v, T$$
$$\stackrel{*}{\Longrightarrow} \widehat{S}.$$

Our new proof is then

$$S \stackrel{*}{\Longrightarrow} u \approx v, T$$
$$\Longrightarrow_{lp} u_1 \approx l'_1, \ldots, u_i \approx l'_i, \ldots, u_n \approx l'_n,$$
$$u[\alpha \leftarrow r_1] \approx v, T$$
$$\Longrightarrow_{lp} u_1 \approx l'_1, \ldots, u_i \approx l'_i[\beta \leftarrow r_2], \ldots, u_n \approx l'_n,$$
$$Q', u[\alpha \leftarrow r_1] \approx v, T$$
$$\stackrel{*}{\Longrightarrow}_{st} u_1 \approx \sigma(l'_1), \ldots, u_i \approx \sigma(l'_i[\beta \leftarrow r_2]), \ldots, u_n \approx \sigma(l'_n),$$
$$S_\sigma, u[\alpha \leftarrow \sigma(r_1)] \approx v, T$$
$$\stackrel{*}{\Longrightarrow} \widehat{S} \cup S_{\sigma'},$$

where Q' is either $l'_i/\beta \approx l_2$ or $l'_i/\beta \approx l_2 \Longrightarrow_{dec} Q'$. In other respects this case is analogous to the previous three.[8]　□

Finally, we can prove the completeness of the $\mathcal{T}$-transformations in the general case.

Theorem 6.2.8 (Completeness of $\mathcal{T}$) Let E be a set of equations, R a rewrite system, and $\succ$ a reduction ordering containing R total on ground terms. Given any finite system S, if θ is an (E, R)-unifier of S, then there is a sequence of $\mathcal{T}$-transformations $S \stackrel{*}{\Longrightarrow} \widehat{S}$ (using variants of equations in $E \cup E^{-1} \cup R \cup R^{-1}$) yielding a solved system $\widehat{S}$ such that if $\sigma_{\widehat{S}}$ is the substitution associated with $\widehat{S}$, then $\sigma_{\widehat{S}} \leq_{E \cup R} \theta[Var(S)]$.

Proof. By Theorem 6.1.7, $E^\omega \cup R^\omega$ is equivalent to (E, R) and is ground Church-Rosser relative to $\succ$. By Lemma 6.2.5, there is a sequence of $\mathcal{T}$-transformations $S \stackrel{*}{\Longrightarrow} \widehat{S}$ using variants of equations in $E^\omega \cup (E^\omega)^{-1} \cup R^\omega$ yielding a solved system $\widehat{S}$ such that if $\sigma_{\widehat{S}}$ is the substitution associated

---

[8] We should remark here that the original proof of this lemma in [54] contained an error which was noticed by Dan Dougherty and Patty Johann [39]; the fact that the definition of lazy paramodulation contained an implicit decomposition step in certain cases was overlooked, and hence the proof of the lemma was insufficient for the claim. See also Section §6.5.

with $\widehat{S}$, then $\sigma_{\widehat{S}} \leq_{E \cup R} \theta[Var(S)]$. Finally, we use Lemma 6.2.7 to eliminate uses of critical pairs, obtaining a sequence where all equations are in $E \cup E^{-1} \cup R \cup R^{-1}$. $\square$

Note that when (E, R) is ground Church-Rosser, equations in E are used as two-way rules in Lazy Paramodulation, but rules in R can be used oriented. This means that in a step

$$u \approx v \Longrightarrow u/\beta \approx l, u[\beta \leftarrow r] \approx v,$$

where β is a nonvariable occurrence in u, then $l \doteq r \in E \cup E^{-1}$ if $l \doteq r$ is not in R, but $r \rightarrow l$ is not tried if $l \rightarrow r$ is in R, and similarly for a step

$$u \approx v \Longrightarrow u \approx v[\beta \leftarrow r], l \approx v/\beta,$$

where β is a nonvariable occurrence in v. Furthermore, Lazy Paramodulation can be restricted so that it applies only when either $\beta = \epsilon$ or one of u, v is a variable (but not both). This is in contrast to the general case where even rules in R may have to be used as two-way rules due to the computation of critical pairs. Also, Lazy Paramodulation may have to be applied with $\beta \neq \epsilon$ even when both u and v are not variables. This case only seems necessary to compute critical pairs. So far, we have failed to produce an example where Lazy Paramodulation needs to be applied in its full generality (that is, when neither u nor v is a variable and $\beta \neq \epsilon$). We conjecture that T is still complete if Lazy Paramodulation is restricted so that it applies only when either $\beta = \epsilon$ or one of u, v is a variable (but not both). The following example might help the reader's intuition.

Example 6.2.9 Let $E = \{fgx \doteq x \ [1], \ ghy \doteq gky \ [2], \ gkfz \doteq z \ [3]\}$, and consider finding E-unifiers for the equation $hu \approx u$. Equations [1] and [2] overlap at 1 in fgx, and we get the critical pair $hv \doteq fgkv$ [4]. We have the sequence of transformations:

$$
\begin{aligned}
hu \approx u \Longrightarrow_{lp} \ & hu \approx hv, fgkv \approx u && \text{using [4]} \\
\Longrightarrow_{lp} \ & hu \approx hv, gkv \approx gkfz, fz \approx u && \text{using [3]} \\
\Longrightarrow_{dec} \ & u \approx v, kv \approx kfz, fz \approx u && \\
\Longrightarrow_{dec} \ & u \approx v, v \approx fz, fz \approx u && \\
\Longrightarrow_{vel} \ & u \approx v, u \approx fz, fz \approx u && \text{applied to } v \\
\Longrightarrow_{vel} \ & fz \approx v, u \approx fz, fz \approx fz && \text{applied to } u \\
\Longrightarrow_{triv} \ & fz \approx v, u \approx fz &&
\end{aligned}
$$

Thus, $[fz/u, fz/v]$ is an E-unifier of $hu \approx u$, and $[fz/u]$ belongs to a complete set of E-unifiers for $hu \approx u$. Interestingly, $[fz/u]$ can also be found using the original equations $[1], [2], [3]$.

$$
\begin{aligned}
hu \approx u \Longrightarrow_{lp}\ & hu \approx x, fgx \approx u && \text{using } [1] \\
\Longrightarrow_{lp}\ & hu \approx x, gx \approx gkfz, fz \approx u && \text{using } [3] \\
\Longrightarrow_{vel}\ & hu \approx x, ghu \approx gkfz, fz \approx u && \text{applied to } x \\
\Longrightarrow_{lp}\ & hu \approx x, ghu \approx ghy, gky \approx gkfz, fz \approx u && \text{using } [2] \\
\overset{*}{\Longrightarrow}_{dec}\ & hu \approx x, u \approx y, y \approx fz, fz \approx u \\
\Longrightarrow_{vel}\ & hu \approx x, u \approx y, u \approx fz, fz \approx u && \text{applied to } y \\
\Longrightarrow_{vel}\ & hfz \approx x, fz \approx y, u \approx fz, fz \approx fz && \text{applied to } u \\
\Longrightarrow_{triv}\ & hfz \approx x, fz \approx y, u \approx fz
\end{aligned}
$$

Thus, $[fz/u, hfz/x, fz/y]$ is an E-unifier of $hu \approx u$.

Lemma 6.2.5 also provides a rigorous proof of the correctness of the transformations of Martelli, Moiso, and Rossi [111] in the case where $E = \emptyset$ and R is canonical. In fact, we have shown the more general case where R is ground Church-Rosser w.r.t. $\succ$.

6.3 Surreduction

An alternate proof of the completeness of the T-transformations is provided in this section by showing that the rewrite steps occurring in a rewrite proof of $\sigma(u) \overset{*}{\longleftrightarrow}_E \sigma(v)$ can be simulated by certain generalizations of rewrite steps called surreduction steps (or narrowing steps). It should be noted that this completeness result is weaker than the completeness results given by Lemma 6.2.5 and Theorem 6.2.8. This point will be clarified in the next section. These results generalize those found in Chapter §4.

Definition 6.3.1 Let E be a set of equations (or a rewrite system) and let W be a set of protected variables. Given any two terms u, v, we say that there is a *surreduction step* (or *narrowing step*) from u to v away from W iff there is some address β in u where u/β is not a variable, a variant $l \doteq r$ of an equation in $E \cup E^{-1}$ (or E if E is a rewrite system) such that u/β and l are unifiable and the variables in $Var(l, r)$ are *new* and occur *only* in l and r (so that $Var(l, r) \cap (Var(u) \cup W) = \emptyset$) and if $\sigma = mgu(u/\beta, l)[W]$, then $v = \sigma(u[\beta \leftarrow r])$. A surreduction step is denoted as

$$
u \ \succ\!\!\longrightarrow_{[\beta, l \doteq r, \sigma, W]} \ v.
$$

(some arguments may be omitted). The substitution σ is called the *surreducing substitution*. A surreduction sequence (or narrowing sequence) is defined as in Chapter §4. Recall that a surreduction step

$$u \succ\!\!\!\longrightarrow_{[\beta,l\doteq r,\sigma]} v$$

corresponds to the rewrite step

$$\sigma(u) \longrightarrow_{[\beta,l\doteq r,\sigma]} v.$$

The crucial lemma in proving the completeness result of this section is a version of the "lifting lemma" from Chapter §4. Since we are not necessarily dealing with rewrite rules ($Var(r)$ is not necessarily a subset of $Var(l)$ for an equation $l \doteq r$), we give a detailed proof of our extension of this result.

Lemma 6.3.2 Let E be a set of equations, R a rewrite system, $\succ$ a reduction ordering containing R, u a term, W a set of 'protected variables' containing $Var(u)$, θ a ground substitution reduced w.r.t. $R(E) \cup R$ such that $D(\theta) \subseteq W$, and $\rho(l) \to \rho(r)$ a ground rule such that either $l \to r$ is a variant of a rule in R or a variant of an equation in E such that $\rho(l) \to \rho(r) \in R(E)$, $D(\rho) = Var(l,r)$ and by the variant assumption, the variables in $Var(l,r)$ are *new* and occur only in this rule. For any ground term v, if

$$\theta(u) \longrightarrow_{[\beta,l\doteq r,\rho]} v,$$

for some address $\beta \in \theta(u)$, then there are two substitutions θ' and σ, a new set of protected variables W', and a term v' such that:

(1) u/β is not a variable and σ is the mgu of u/β and l away from $W \cup Var(l,r)$

(2) $v' = \sigma(u)[\beta \leftarrow \sigma(r)]$ and $\sigma(u) \longrightarrow_{[\beta,l\doteq r,\sigma]} v'$

(3) $\theta = \sigma \circ \theta'[W]$ and $\theta'|_{W \cup I(\sigma)}$ is reduced w.r.t. $R(E) \cup R$

(4) $v = \theta'(v')$ and

(5) $Var(v') \subseteq W'$ and $D(\theta') \subseteq W'$.

This may be illustrated as follows:

$$
\begin{array}{ccc}
\theta(u) & \longrightarrow_{[\beta,l\doteq r,\rho]} & v = \theta'(v') \\
{\scriptstyle\theta}\uparrow & & \uparrow{\scriptstyle\theta'} \\
u & \succ\!\!\!\longrightarrow_{[\beta,l\doteq r,\sigma,W]} & v'
\end{array}
$$

Proof. Obviously, $\theta(u)/\beta = \rho(l)$. Since θ is reduced w.r.t. $R(E) \cup R$, β must be the address of a nonvariable symbol in u, and $\theta(u)/\beta = \theta(u/\beta)$. Let $t = u/\beta$. Since $D(\theta) \cap D(\rho) = \emptyset$, we can form the union $\varphi = \theta \cup \rho$ of the substitutions θ and ρ, and we have $\varphi(t) = \varphi(l)$, i.e., φ is a unifier of t and l. By Lemma 3.3.11 we have an mgu σ of t and l away from $W \cup Var(l, r)$, proving (1). Also, by corollary 3.3.12 there is some substitution η such that $\varphi = \theta \cup \rho = \sigma \circ \eta[W \cup Var(l)]$, where w.l.g., since σ is idempotent, we can assume that $D(\eta) \cap D(\sigma) = \emptyset$. Also note that since $Var(l)$ and $Var(u)$ are disjoint, then $D(\sigma) = Var(t) \cup Var(l)$. Let $v' = \sigma(u)[\beta \leftarrow \sigma(r)]$. Observe that the variables in v' are contained in the union of the three disjoint sets W, $I(\sigma)$, and $(Var(r) - Var(l))$. This last set is nonempty when $Var(r)$ is not a subset of $Var(l)$, which is possible when $\rho(l) \doteq \rho(r)$ is an orientable instance. We define $W' = W \cup I(\sigma) \cup (Var(r) - Var(l))$ (proving the first part of (5)), and we define the substitution θ' as follows:

$$\theta'(y) = \begin{cases} \eta(y), & \text{if } y \in W \cup I(\sigma); \\ \rho(y), & \text{if } y \in (Var(r) - Var(l)). \end{cases}$$

Clearly, the first part of (5) holds. Since $v' = \sigma(u)[\beta \leftarrow \sigma(r)]$ and $\sigma(u)/\beta = \sigma(t) = \sigma(l)$ (because σ is a unifier of t and l), we have

$$\sigma(u) \quad \longrightarrow_{[\beta, l \doteq r, \sigma]} \quad v'$$

and (2) holds. Since

$$\theta(u) \quad \longrightarrow_{[\beta, l \doteq r, \rho]} \quad v,$$

we have $v = \theta(u)[\beta \leftarrow \rho(r)]$. We now show that $v = \theta'(v')$. Since $v' = \sigma(u)[\beta \leftarrow \sigma(r)]$, we have $\theta'(v') = \theta'(\sigma(u))[\beta \leftarrow \theta'(\sigma(r))]$. Hence, we need to show that

$$\theta'(\sigma(u))[\beta \leftarrow \theta'(\sigma(r))] = \theta(u)[\beta \leftarrow \rho(r)].$$

Since $\theta \cup \rho = \sigma \circ \eta[W \cup Var(l)]$ and $\theta' = \eta[W \cup I(\sigma)]$, then by the definition of θ' and the variant assumption we have $\theta = \sigma \circ \theta'[W]$ and $\theta'(\sigma(u)) = \theta(u)$. This also shows the first part of (3). Since $\theta \cup \rho = \sigma \circ \eta[W \cup Var(l)]$ and $\theta' = \eta[W \cup I(\sigma)]$, if $y \in Var(l) \cap Var(r)$, then $\theta'(\sigma(y)) = \rho(y)$. If $y \in Var(r) - Var(l)$, since $\theta'(y) = \rho(y)$ and $\sigma(y) = y$ (because $D(\sigma) = Var(l) \cup Var(t)$), we also have $\theta'(\sigma(y)) = \rho(y)$. Hence, $\theta'(\sigma(r)) = \rho(r)$, and we have shown that $v = \theta'(v')$. Thus, (4) holds. It remains to show the second part of (3), that $\theta'|_{W \cup I(\sigma)}$ is reduced w.r.t. $R(E) \cup R$. Recall that $\theta' = \eta[W \cup I(\sigma)]$. Thus, we show that η is reduced w.r.t. $R(E) \cup R$ on $W \cup I(\sigma)$. For any $y \in D(\eta) \cap (W \cup I(\sigma))$, there are two cases. If $y \in W$, then, since $D(\theta') \cap D(\sigma) = \emptyset$, $\sigma(y) = y$, and since $\theta \cup \rho = \sigma \circ \eta[W \cup Var(l)]$,

$\eta(y) = \eta(\sigma(y)) = \theta(y)$. Since $\theta(y)$ is reduced w.r.t. $R(E) \cup R$, so is $\eta(y)$. Now by the definition of σ and by the variant assumption, we have $I(\sigma) = Var(\sigma(t))$ and $Var(\sigma(t)) \cap Var(t) = \emptyset$. Also, since $\theta \cup \rho = \sigma \circ \eta[W \cup Var(l)]$, then for every variable z in $Var(t)$, $\theta(z) = \eta(\sigma(z))$. Hence, for every $y \in I(\sigma)$, $\eta(y) = \theta(z)/\alpha$ for some $z \in Var(t)$, where α is the address of y in $\sigma(z)$. Since $\theta(z)$ is reduced w.r.t. $R(E) \cup R$, so is its subterm $\eta(y)$. Thus (3) holds, and the proof is complete. □

We now have the following result showing the crucial role played by surreductions.

Lemma 6.3.3 Let E be a set of equations, R a rewrite system, and $\succ$ a reduction ordering containing R, and assume that (E, R) is ground Church-Rosser relative to $\succ$. Let the symbol eq be a new binary function symbol not in Σ. Given any two terms u, v, if a ground substitution θ reduced w.r.t. $R(E) \cup R$ and such that $Var(u, v) \subseteq D(\theta)$ is an (E, R)-unifier of u and v, then for any set of protected variables W containing $D(\theta)$, there is a surreduction sequence

$$eq(u, v) \ \succ\!\!\longrightarrow_{[l_1 \doteq r_1, \sigma_1]} \ eq(u_1, v_1) \ldots \ \succ\!\!\longrightarrow_{[l_n \doteq r_n, \sigma_n]} \ eq(u_n, v_n)$$

(where each $l_i \doteq r_i$ is a variant of an equation in $E \cup E^{-1} \cup R$) and some *mgu* μ of u_n and v_n such that

$$\sigma_1 \circ \ldots \circ \sigma_n \circ \mu \leq \theta[W].$$

Furthermore, the substitution $\sigma_1 \circ \ldots \circ \sigma_n \circ \mu|_{Var(u,v)}$ is an (E, R)-unifier of u and v.

Proof. Since (E, R) is ground Church-Rosser relative to $\succ$, there is a rewrite proof

$$\theta(u) \ \xrightarrow{\ *\ }_{R(E) \cup R} \ N \ \xleftarrow{\ *\ }_{R(E) \cup R} \ \theta(v),$$

where N is irreducible (w.r.t. $R(E) \cup R$). Hence, there is a rewrite proof

$$\theta(eq(u, v)) \ \xrightarrow{\ *\ }_{R(E) \cup R} \ eq(N, N),$$

where $eq(N, N)$ is irreducible. We proceed by induction on the well-founded ordering $\succ$. If $\theta(eq(u, v))$ is irreducible, obviously $eq(\theta(u), \theta(v)) = eq(N, N)$, and θ is a unifier of u and v. The lemma is satisfied by choosing μ as a $mgu(u, v)[W]$. Otherwise, there is a rewrite proof

$$\theta(eq(u, v)) \ \longrightarrow_{[\beta, l \doteq r, \rho]} \ w \ \xrightarrow{\ *\ }_{R(E) \cup R} \ eq(N, N),$$

where $\rho(l) \rightarrow \rho(r) \in R(E)$ or $l \rightarrow r \in R$, and $\rho(l) \succ \rho(r)$. If $\rho|_{Var(r)-Var(l)}$ is not reduced, since $R(E) \cup R$ is canonical on ground terms, we can reduce each $\rho(x)$ where $x \in Var(r) - Var(l)$ to its normal form $\rho(x) \downarrow$ (w.r.t. $R(E) \cup R$), obtaining a reduced substitution ρ_1. But then, using the rule $\rho_1(l) \rightarrow \rho_1(r)$ which also satisfies $\rho_1(l) \succ \rho_1(r)$, since $\rho|_{Var(l)} = \rho_1|_{Var(l)}$ and $\rho(y) \succ \rho_1(y)$ for each $y \in Var(r) - Var(l)$, we have a rewrite proof

$$\theta(eq(u,v)) \quad \longrightarrow_{[\beta, l \doteq r, \rho_1]} \quad w_1 \quad \overset{*}{\longrightarrow}_{R(E) \cup R} \quad eq(N,N).$$

Then, by Lemma 6.3.2, we have a surreduction step away from W

$$eq(u,v) \quad \succ\!\!\!\longrightarrow_{[\beta, l \doteq r, \sigma_1, W]} \quad w_1',$$

substitutions σ_1 and θ_1, and $W' = W \cup I(\sigma) \cup (Var(r) - Var(l))$ such that $\theta_1(w_1') = w_1$, $\theta = \sigma_1 \circ \theta_1[W]$, $D(\theta_1)$, $Var(w_1') \subseteq W'$, and the substitution $\theta_1|_{W \cup I(\sigma)}$ is reduced w.r.t. $R(E) \cup R$. Since $\theta_1|_{Var(r)-Var(l)} = \rho_1|_{Var(r)-Var(l)}$ and ρ_1 is reduced (w.r.t. $R(E) \cup R$), θ_1 is reduced w.r.t. $R(E) \cup R$. But w_1' is of the form $eq(u_1, v_1)$ and $w_1 = \theta_1(eq(u_1, v_1))$. Also, since $\rho_1(l) \succ \rho_1(r)$ and

$$\theta(eq(u,v)) \quad \longrightarrow_{[\beta, l \doteq r, \rho_1]} \quad \theta_1(eq(u_1, v_1)),$$

we have $\theta(eq(u,v)) \succ \theta_1(eq(u_1, v_1))$. Since $w_1 \overset{*}{\longrightarrow}_{R(E) \cup R} eq(N,N)$, we have

$$\theta_1(eq(u_1, v_1)) \overset{*}{\longrightarrow}_{R(E) \cup R} eq(N,N).$$

Hence, the induction hypothesis applies using the new set of protected vars $W' = W \cup I(\sigma) \cup (Var(r) - Var(l))$, and there is some surreduction sequence

$$eq(u_1, v_1) \succ\!\!\!\longrightarrow_{[l_2 \doteq r_2, \sigma_2]} eq(u_2, v_2) \ldots \succ\!\!\!\longrightarrow_{[l_n \doteq r_n, \sigma_n]} eq(u_n, v_n)$$

and some mgu μ of u_n and v_n such that

$$\sigma_2 \circ \ldots \circ \sigma_n \circ \mu \leq \theta_1[W'].$$

Since $\theta = \sigma_1 \circ \theta_1[W]$, we have

$$\sigma_1 \circ \ldots \circ \sigma_n \circ \mu \leq \theta[W].$$

The proof that $\sigma_1 \circ \ldots \circ \sigma_n \circ \mu|_{Var(u,v)}$ is an (E,R)-unifier of u and v is an (E,R)-unifier of u and v is a routine extension of Lemma 4.2.6 $\square$

The previous lemma implies the following important theorem.

Theorem 6.3.4 Let E be a set of equations, R a rewrite system, and $\succ$ a reduction ordering containing R total on ground terms. Given any two terms u, v, if θ is an (E, R)-unifier of u and v, then for any set W containing $Var(u, v)$ and $D(\theta)$ there is a surreduction sequence

$$eq(u, v) \succ\!\!\!\longrightarrow_{[l_1 \doteq r_1, \sigma_1]} eq(u_1, v_1) \ldots \succ\!\!\!\longrightarrow_{[l_n \doteq r_n, \sigma_n]} eq(u_n, u_n)$$

(where each $l_i \doteq r_i$ is a variant of an equation in $E^\omega \cup (E^\omega)^{-1} \cup R^\omega$) and a mgu μ of u_n and v_n such that

$$\sigma_1 \circ \ldots \circ \sigma_n \circ \mu \leq_{EUR} \theta[W].$$

Furthermore, $\sigma_1 \circ \ldots \circ \sigma_n \circ \mu|_{Var(u,v)}$ is an (E, R)-unifier of u and v.

Proof. First, recall that by Lemma 6.1.9 it can be assumed that θ is ground and that $Var(u, v) \subseteq D(\theta)$ without any loss of generality. Next, we use Theorem 6.1.7 which shows that $E^\omega \cup R^\omega$ is equivalent to (E, R) and is ground Church-Rosser relative to $\succ$. Then, by Lemma 6.1.10, we know that there is a ground substitution θ'' reduced w.r.t. $R(E^\omega) \cup R^\omega$ and such that $\theta'' =_{EUR} \theta'[Var(u, v)]$. Finally, we apply Lemma 6.3.3 to θ'' and $R(E^\omega) \cup R^\omega$. $\square$

It is remarkable that Theorem 6.3.4 shows the completeness of surreduction together with the computation of critical pairs. Note that rules in R^ω can be applied oriented, whereas equations in E^ω have to be used as two-way rules. This adds considerably to the nondeterminism of the method, and shows why oriented rules are preferred. We now show how a weaker version of the completeness of our T-transformations can be obtained from Theorem 6.3.4.

6.4 Completeness of T Revisited

First, we show that the T-transformations can simulate surreduction in the case of a pair (E, R) that is ground Church-Rosser (w.r.t. $\succ$).

Lemma 6.4.1 Let E be a set of equations, R a rewrite system, and $\succ$ a reduction ordering containing R. Assume that (E, R) is ground Church-Rosser (w.r.t. $\succ$). For every surreduction sequence

$$eq(u, v) \succ\!\!\!\longrightarrow_{[l_1 \doteq r_1, \sigma_1]} eq(u_1, v_1) \ldots \succ\!\!\!\longrightarrow_{[l_n \doteq r_n, \sigma_n]} eq(u_n, v_n)$$

where each $l_i \doteq r_i$ is a variant of an equation in $E \cup E^{-1} \cup R$ and μ is the mgu of u_n and v_n, there is a sequence of T-transformations $u \approx v \stackrel{*}{\Longrightarrow} S$ yielding a solved system S such that

$$\sigma_S = \sigma_1 \circ \ldots \circ \sigma_n \circ \mu[Var(u, v)].$$

Proof. The lemma is proved by induction on the length of surreduction sequences. If $n = 0$, then u and v are unifiable by μ, and by the completeness of the transformations for standard unification (without Lazy Paramodulation), the result holds. Otherwise, since $eq(u, v) \succ\!\!\rightarrow_{[\sigma_1]} eq(u_1, v_1)$, either

$$u \;\succ\!\!\rightarrow_{[\beta, l \doteq r, \sigma_1]}\; u_1$$

for some address β in u and $v_1 = \sigma_1(v)$, or $u_1 = \sigma_1(u)$ and

$$v \;\succ\!\!\rightarrow_{[\beta, l \doteq r, \sigma_1]}\; v_1$$

for some address β in v. We consider the first case, the other being similar. By the induction hypothesis, $u_1 \approx v_1 \overset{*}{\Longrightarrow} S'$ by a sequence of T-transformations, where S' is a solved system such that

$$\sigma_{S'} = \sigma_2 \circ \ldots \circ \sigma_n \circ \mu[Var(u, v)].$$

However, since $eq(u, v) \succ\!\!\rightarrow_{[\beta, l \doteq r, \sigma_1]} eq(u_1, v_1)$, we have

$$u \approx v \Longrightarrow u/\beta \approx l, u[\beta \leftarrow r] \approx v$$

by Lazy Paramodulation, and

$$u/\beta \approx l, u[\beta \leftarrow r] \approx v \overset{*}{\Longrightarrow} S_1 \cup \sigma_1(u[\beta \leftarrow r]), \sigma_1(v) = S_1 \cup u_1, v_1,$$

by performing the sequence of transformations from the set ST that computes the mgu σ_1 of u/β and l and the corresponding solved system S_1. Thus,

$$u \approx v \overset{*}{\Longrightarrow} S_1 \cup u_1 \approx v_1.$$

Since by the induction hypothesis

$$u_1 \approx v_1 \overset{*}{\Longrightarrow} S',$$

it is easy to see (by induction on the length of the sequence) that

$$S_1 \cup u_1 \approx v_1 \overset{*}{\Longrightarrow} \sigma_{S'}(S_1) \cup S',$$

and so

$$u \approx v \overset{*}{\Longrightarrow} \sigma_{S'}(S_1) \cup S',$$

and letting $S = \sigma_{S'}(S_1) \cup S'$, S is in solved form. Since S_1 is the system in solved form associated with σ_1, and since the substitutions σ_i and μ have pairwise disjoint domains, we have

$$\sigma_S = \sigma_1 \circ \ldots \circ \sigma_n \circ \mu[Var(u, v)]. \qquad \square$$

We can now give another proof of the completeness of the set of transformations T when (E, R) is ground Church-Rosser.

Lemma 6.4.2 Let E be a set of equations, R a rewrite system, and $\succ$ a reduction ordering containing R. The set of transformations $\mathcal{T}$ is complete for all ground Church-Rosser pairs (E, R).

Proof. We need to prove that given any two terms u, v, if θ is an (E, R)-unifier of u and v, then there is a sequence of $\mathcal{T}$-transformations $u \approx v \stackrel{*}{\Longrightarrow} S$ (using variants of equations in $E \cup E^{-1} \cup R$) yielding a solved system S such that if σ_S is the substitution associated with S, then $\sigma_S \leq_{EUR} \theta[Var(u, v)]$. Without loss of generality, by Lemma 6.1.9, it can be assumed that θ is ground and that $Var(u, v) \subseteq D(\theta)$. By Lemma 6.1.10, there is a ground substitution θ' reduced w.r.t. $R(E) \cup R$ and such that $\theta' =_{EUR} \theta[Var(u, v)]$. By Lemma 6.3.3, there is a surreduction sequence

$$eq(u, v) \succ\!\!\!\rightarrow_{[l_1 \doteq r_1, \sigma_1]} eq(u_1, v_1) \ldots \succ\!\!\!\rightarrow_{[l_n \doteq r_n, \sigma_n]} eq(u_n, v_n)$$

where each $l_i \doteq r_i$ is a variant of an equation in $E \cup E^{-1} \cup R$, u_n and v_n are unifiable, and if μ is the mgu of u_n and v_n, then

$$\sigma_1 \circ \ldots \circ \sigma_n \circ \mu \leq \theta'[Var(u, v)].$$

By Lemma 6.4.1, there is a sequence of $\mathcal{T}$-transformations $u \approx v \stackrel{*}{\Longrightarrow} S$ yielding a solved system S such that

$$\sigma_S = \sigma_1 \circ \ldots \circ \sigma_n \circ \mu[Var(u, v)].$$

Thus,

$$\sigma_S = \sigma_1 \circ \ldots \circ \sigma_n \circ \mu \leq \theta' =_{EUR} \theta[Var(u, v)],$$

and so $\sigma_S \leq_{EUR} \theta[Var(u, v)]$. $\square$

It is worth noting that Lemma 6.4.2 is weaker than Lemma 6.2.5 in the following sense. Lemma 6.2.5 shows the completeness of the transformations $\mathcal{T}$ even when Lazy Paramodulation is restricted to apply either at the top ($\beta = \epsilon$) or when one of u, v is a variable (but not both). However, this is not the case for Lemma 6.4.2. The simulation of surreduction steps requires Lazy Paramodulation unrestricted. This is not very surprising. In the proof of Lemma 6.2.5, transformations are applied in a *top-down* and lazy fashion. By lazy, we mean that unification steps can be delayed. On the other hand, it is not clear that completeness is guaranteed if such a top-down strategy is applied in a sequence of surreduction steps. However, using Lemma 6.1.12, it can be shown that surreduction steps can always be applied bottom-up, that is, using innermost steps, and it is easy to see that Lemma 6.4.2 still holds under this strategy. This corresponds to a *bottom-up* strategy for applying the transformations, and the proof of Lemma 6.2.5 does not yield the completeness of this strategy. Thus, it appears that Lemma 6.2.5 and Lemma 6.4.2 correspond to different strategies for applying the transformations, and that they are complementary.

In a recent paper, Nutt, Réty, and Smolka [120] investigate complete sets of transformations for basic narrowing applied to ground confluent systems. It would be interesting to explore the relationship between our set of transformations T and the transformations presented in [120].

Finally, we give an alternate proof of the completeness of the set T in the general case. The above comments also apply to this theorem and to Theorem 6.2.8.

Theorem 6.4.3 Let E be a set of equations, R a rewrite system, and $\succ$ a reduction ordering containing R total on ground terms. The set T is a complete set of transformations.

Proof. Without loss of generality, we can assume that θ is ground and that $Var(u,v) \subseteq D(\theta)$. By Theorem 6.1.7, $E^\omega \cup R^\omega$ is equivalent to (E, R) and is ground Church-Rosser relative to $\succ$. Then, by Lemma 6.4.2, there is a sequence of T-transformations $u \approx v \stackrel{*}{\Longrightarrow} S$ using equations in $E^\omega \cup (E^\omega)^{-1} \cup R^\omega$ yielding a solved system S such that $\sigma_S \leq_{E \cup R} \theta[Var(u,v)]$, where σ_S is the substitution associated with S. We conclude by applying Lemma 6.2.7. $\square$

6.5 Relaxed Paramodulation

It is possible to improve the set of transformations T further by strengthening the restriction on the Lazy Paramodulation rule, as shown recently by D. Dougherty and P. Johann [39]. The basic idea is that the restriction that u/α and l_1 have identical top function symbols when l_1 is not a variable, and that decomposition is applied to the new equation as a part of the rule, can be iterated down into the two terms. We may formalize this new form of the Lazy Paramodulation rule by defining a new form of partial unification as follows.

Definition 6.5.1 (Top Unification) Given an equation $s \approx t$, $TU(s \approx t)$ is a set of equations defined recursively:

$$TU(f(\ldots) \approx g(\ldots)) \text{ is } undefined \text{ if } f \neq g;$$
$$TU(x \approx t) = \{x \approx t\};$$

$$TU\big(f(s_1, \ldots, s_n) \approx f(t_1, \ldots, t_n)\big) = \bigcup_{1 \le i \le n} TU(s_i \approx t_i),$$

for $n \ge 0$, provided these last are defined; otherwise the result is undefined. Two terms s and t are said to *top unify* if $TU(s \approx t)$ is defined.

The idea is, informally, to decompose an equation down completely into variable–term or term–variable equations; that is, since decomposition is a terminating relation, we normalize an equation $s \approx t$ wrt decomposition,

$$s \approx t \implies_{dec} \implies_{dec} \cdots \implies_{dec} \{s_1 \approx t_1, \ldots, s_n \approx t_n\}$$

where no decomposition can be applied to any $s_i \approx t_i$. Then $s \approx t$ is top unifiable iff at least one of s_i, t_i is a variable, for each i, $1 \le i \le n$.

For example

$$TU\big(f(c, x, g(x, a)) \approx f(c, h(x), g(y, a))\big) = \{x \approx h(x), x \approx y\},$$

but $TU\big(f(c, x, g(x, a)) \approx f(d, h(x), g(y, a))\big)$ is undefined. Note that top unifiability does not imply unifiability, since perhaps in the second line of the definition $x \in Var(t)$, but of course unifiability implies top unifiability.

Before we discuss the E-unification procedure based on this partial form of unification, we give a collection of results related to top unification which will be necessary below. Most of these are simple properties of the relation $\implies_{dec}$, and their proofs are left to the reader (see also [39]).

Proposition 6.5.2 Let $s \approx t$ be an equation.

(1) If $\sigma = mgu(s, t)$, then there exists a sequence

$$s \approx t \overset{*}{\implies}_{dec} TU(s \approx t) \overset{*}{\implies}_{st} S_\sigma.$$

(2) For any decomposition step $s \approx t \implies_{dec} T$ and any substitution σ, there exists a decomposition step $\sigma(s \approx t) \implies_{dec} \sigma(T)$.

(3) If σ is a substitution such that $\sigma(s \approx t)$ is top unifiable, then there exists a sequence

$$\sigma(s \approx t) \overset{*}{\implies}_{dec} \sigma(TU(s \approx t)) \overset{*}{\implies}_{dec} TU(\sigma(s \approx t)).$$

(4) If $\beta \in Dom(s) \cap Dom(t)$, and $s \approx t$ is top unifiable, then $TU(s \approx t) = TU(s/\beta \approx t/\beta) \cup T'$ for some T'. Furthermore, if $s' = s[\beta\gamma \leftarrow u]$, then $TU(s' \doteq t) = TU(s/\beta[\gamma \leftarrow u] \approx t/\beta) \cup T'$

Note that (3) implies that if $\sigma(s \approx t)$ is top unifiable, then so is $s \approx t$. The new method for E-unification based on this form of partial unification may be defined.

Definition 6.5.3 (The set of transformations T') The set T' consists of the rules from $\mathcal{ST}$ plus:

Relaxed Paramodulation: Given a pair (E, R) and a system $\{s \approx t\} \cup S$,

$$\{s \approx t\} \cup S \Longrightarrow TU(s/\beta \approx l) \cup \{s[\beta \leftarrow r] \approx t\} \cup S$$

if β is a non-variable occurrence in s, $l \doteq r$ is a variant from $E \cup E^{-1} \cup R \cup R^{-1}$, and $TU(s/\beta \approx l)$ is defined.

The completeness of this set is proved by strengthening Lemma 6.2.7 to accomodate this stronger restriction; the rest of the proof is essentially unchanged from Section §6.2 (for an alternate proof see [39]).

For the purposes of the proof below it will be convenient to consider the Relaxed Paramodulation rule in two parts, namely, a "Completely Lazy Paramodulation" transformation

$$\{s \approx t\} \cup S \Longrightarrow_{clp} \{s/\beta \approx l, s[\beta \leftarrow r] \approx t\} \cup S$$

followed by a sequence of decomposition steps

$$\{s/\beta \approx l, s[\beta \leftarrow r] \approx t\} \cup S$$
$$\overset{*}{\Longrightarrow}_{dec} TU(s/\beta \approx l) \cup \{s[\beta \leftarrow r] \approx t\} \cup S.$$

The new form of Lemma 6.2.7 we need will proceed by showing how to manipulate $\Longrightarrow_{clp}$ and $\Longrightarrow_{dec}$ to transform the proofs to accomodate this more restrictive form of lazy paramodulation.

Lemma 6.5.4 Let E be a set of equations, R a rewrite system, and $\succ$ a reduction ordering containing R. For every finite system S, every sequence of T'-transformations $S \overset{*}{\Longrightarrow} \widehat{S}$ using equations in $E^\omega \cup (E^\omega)^{-1} \cup R^\omega$ can be converted into a sequence $S \overset{*}{\Longrightarrow} \widehat{S}'$ using equations only in $E \cup E^{-1} \cup R \cup R^{-1}$, such that $\widehat{S}$ and $\widehat{S}'$ are in solved form and $\sigma_{\widehat{S}}|_{Var(S)} = \sigma_{\widehat{S}'}|_{Var(S)}$.

Proof. We again proceed by induction on the (finite) multiset of the depths of all clauses used in the proof $S \overset{*}{\Longrightarrow} \widehat{S}$. The base case is again trivial; now suppose the proof uses some equation $\sigma(r_1) \doteq \sigma(l_1[\beta \leftarrow r_2])$ of depth greater than 0, obtained by forming a critical pair from $l_1 \doteq r_1$ and $l_2 \doteq r_2$ at β in l_1, with mgu σ of l_1/β and l_2.

Case One. Suppose that $\sigma(r_1)$ is not a variable and the Lazy Paramodulation step is from $\sigma(r_1)$.[9] In this case our proof must have the form

$$S \overset{*}{\Longrightarrow} u \approx v, T$$
$$\Longrightarrow_{clp} u/\alpha \approx \sigma(r_1), u[\alpha \leftarrow \sigma(l_1[\beta \leftarrow r_2])] \approx v, T$$
$$\overset{*}{\Longrightarrow}_{dec} TU(u/\alpha \approx \sigma(r_1)), u[\alpha \leftarrow \sigma(l_1[\beta \leftarrow r_2])] \approx v, T$$
$$\overset{*}{\Longrightarrow} \widehat{S}$$

---

[9] This combines cases one through three of Lemma 6.2.7.

where u/α and l_2/β are non-variables.

The new proof is

$$S \overset{*}{\Longrightarrow} u \approx v, T$$
$$\Longrightarrow_{clp} u/\alpha \approx r_1, u[\alpha \leftarrow l_1] \approx v, T$$
$$\overset{*}{\Longrightarrow}_{dec} TU(u/\alpha \approx r_1), u[\alpha \leftarrow l_1] \approx v, T$$
$$\Longrightarrow_{clp} TU(u/\alpha \approx r_1), l_1/\beta \approx l_2, u[\alpha \leftarrow l_1[\beta \leftarrow r_2]] \approx v, T$$
$$\overset{*}{\Longrightarrow}_{dec} TU(u/\alpha \approx r_1), TU(l_1/\beta \approx l_2), u[\alpha \leftarrow l_1[\beta \leftarrow r_2]] \approx v, T$$
$$\overset{*}{\Longrightarrow}_{st} \sigma(TU(u/\alpha \approx r_1)), S_\sigma, u[\alpha \leftarrow \sigma(l_1[\beta \leftarrow r_2])] \approx v, T$$
$$\overset{*}{\Longrightarrow}_{dec} TU(u/\alpha \approx \sigma(r_1)), S_\sigma, u[\alpha \leftarrow \sigma(l_1[\beta \leftarrow r_2])] \approx v, T$$
$$\overset{*}{\Longrightarrow} S_{\sigma'} \cup \widehat{S},$$

where we have used Proposition 6.5.2 (1) in lines 4 – 6 and part (3) in passing from line 6 to line 7. This is clearly a sequence in the requisite form (in other respects the proof is treated analogously with Lemma 6.2.7).

Case Two. Now suppose that the Lazy Paramodulation step proceeds from $\sigma(l_1[\beta \leftarrow r_2])$, namely, the proof has the form

$$S \overset{*}{\Longrightarrow} u \approx v, T$$
$$\Longrightarrow_{clp} u/\alpha \approx \sigma(l_1[\beta \leftarrow r_2]), u[\alpha \leftarrow \sigma(r_1)] \approx v, T$$
$$\overset{*}{\Longrightarrow}_{dec} TU(u/\alpha \approx \sigma(l_1[\beta \leftarrow r_2])), u[\alpha \leftarrow \sigma(r_1)] \approx v, T$$
$$\overset{*}{\Longrightarrow} \widehat{S}.$$

There are two subcases. First, suppose that β is *not* a non-variable occurrence in u/α; since $u/\alpha \approx \sigma(l_1[\beta \leftarrow r_2])$ is top unifiable, then so is $u/\alpha \approx l_1[\beta \leftarrow r_2]$, and thus there is a unique variable $x = u/\alpha\beta_1$ where $\beta = \beta_1\beta_2$ for some β_2. Now if $t = l_1/\beta_1$ (so that $t/\beta_2 = l_1/\beta$, and hence $\sigma = mgu(t/\beta_2, l_2)$), then by Proposition 6.5.2 (3) and (4), the sequence

$$u/\alpha \approx \sigma(l_1[\beta \leftarrow r_2]) \overset{*}{\Longrightarrow}_{dec} TU(u/\alpha \approx \sigma(l_1[\beta \leftarrow r_2])) \qquad (*)$$

can be arranged into the form

$$u/\alpha \approx \sigma(l_1[\beta \leftarrow r_2]) \overset{*}{\Longrightarrow}_{dec} \sigma(TU(u/\alpha \approx l_1[\beta \leftarrow r_2]))$$
$$= \sigma(x \approx t[\beta_2 \leftarrow r_2], T')$$
$$= x \approx \sigma(t[\beta_2 \leftarrow r_2]), \sigma(T')$$
$$\overset{*}{\Longrightarrow}_{dec} TU(u/\alpha \approx \sigma(l_1[\beta \leftarrow r_2])),$$

for some T'. Furthermore, we know that $TU(u/\alpha \approx l_1) = \{x \approx t\} \cup T'$. Our new proof uses these intermediate forms as follows:

$$S \overset{*}{\Longrightarrow} u \approx v, T$$

$$\Longrightarrow_{clp} u/\alpha \approx l_1, u[\alpha \leftarrow r_1] \approx v, T$$
$$\overset{*}{\Longrightarrow}_{dec} TU(u/\alpha \approx l_1), u[\alpha \leftarrow r_1] \approx v, T$$
$$= \quad x \approx t, T', u[\alpha \leftarrow r_1] \approx v, T$$
$$\Longrightarrow_{clp} t/\beta_2 \approx l_2, x \approx t[\beta_2 \leftarrow r_2], T', u[\alpha \leftarrow r_1] \approx v, T$$
$$\overset{*}{\Longrightarrow}_{dec} TU(t/\beta_2 \approx l_2), x \approx t[\beta_2 \leftarrow r_2], T', u[\alpha \leftarrow r_1] \approx v, T$$
$$\overset{*}{\Longrightarrow}_{st} S_\sigma, x \approx \sigma(t[\beta_2 \leftarrow r_2]), \sigma(T'), u[\alpha \leftarrow \sigma(r_1)] \approx v, T$$
$$\overset{*}{\Longrightarrow}_{dec} S_\sigma, TU(u/\alpha \approx \sigma(l_1[\beta \leftarrow r_2])), u[\alpha \leftarrow \sigma(r_1)] \approx v, T$$
$$\overset{*}{\Longrightarrow} S_{\sigma'} \cup \widehat{S}.$$

This is the required sequence, and the other details are as in Lemma 6.2.7.

Second, suppose β *is* a non-variable occurrence in u/α. In this case, first consider the sequence

$$u/\alpha \approx l_1[\beta \leftarrow r_2] \overset{*}{\Longrightarrow}_{dec} u/\alpha\beta \approx r_2, T''$$
$$\overset{*}{\Longrightarrow}_{dec} TU(u/\alpha\beta \approx r_2), TU(T'')$$
$$= \quad TU(u/\alpha \approx l_1[\beta \leftarrow r_2]),$$

which we know to exist by Proposition 6.5.2 parts (3) and (4). By part (4) of the same proposition, we also have a sequence

$$u/\alpha[\beta \leftarrow l_2] \approx l_1 \overset{*}{\Longrightarrow}_{dec} l_2 \approx l_1/\beta, T''$$
$$\overset{*}{\Longrightarrow}_{dec} TU(l_2 \approx l_1/\beta), TU(T'').$$
$$= \quad TU(u/\alpha[\beta \leftarrow l_2] \approx l_1).$$

(The point here is that T'' can be assumed to be identical in both sequences.) Now by considering the effect of σ on the first of these, we can rearrange the sequence (∗) in the previous paragraph into the form

$$u/\alpha \approx \sigma(l_1[\beta \leftarrow r_2]) \overset{*}{\Longrightarrow}_{dec} u/\alpha\beta \approx \sigma(r_2), \sigma(T'')$$
$$\overset{*}{\Longrightarrow}_{dec} \sigma(TU(u/\alpha\beta \approx r_2)), \sigma(TU(T''))$$
$$\overset{*}{\Longrightarrow}_{dec} TU(u/\alpha\beta \approx \sigma(r_2)), TU(\sigma(T''))$$
$$= \quad TU(u/\alpha \approx \sigma(l_1[\beta \leftarrow r_2])).$$

Then we can again use these intermediate forms to obtain the new proof:

$$S \overset{*}{\Longrightarrow} u \approx v, T$$
$$\Longrightarrow_{clp} u/\alpha\beta \approx r_2, u[\alpha\beta \leftarrow l_2] \approx v, T$$
$$\overset{*}{\Longrightarrow}_{dec} TU(u/\alpha\beta \approx r_2), u[\alpha\beta \leftarrow l_2] \approx v, T$$
$$\Longrightarrow_{clp} TU(u/\alpha\beta \approx r_2), (u[\alpha\beta \leftarrow l_2])/\alpha \approx l_1, (u[\alpha\beta \leftarrow l_2])[\alpha \leftarrow r_1] \approx v, T$$

$$
\begin{aligned}
&= && TU(u/\alpha\beta \approx r_2), u/\alpha[\beta \leftarrow l_2] \approx l_1, u[\alpha \leftarrow r_1] \approx v, T \\
&\stackrel{*}{\Longrightarrow}_{dec} && TU(u/\alpha\beta \approx r_2), TU(u/\alpha[\beta \leftarrow l_2] \approx l_1), u[\alpha \leftarrow r_1] \approx v, T \\
&= && TU(u/\alpha\beta \approx r_2), TU(l_2 \approx l_1/\beta), TU(T''), u[\alpha \leftarrow r_1] \approx v, T \\
&\stackrel{*}{\Longrightarrow}_{st} && \sigma(TU(u/\alpha\beta \approx r_2)), S_\sigma, \sigma(TU(T'')), u[\alpha \leftarrow \sigma(r_1)] \approx v, T \\
&\stackrel{*}{\Longrightarrow}_{dec} && TU(u/\alpha\beta \approx \sigma(r_2)), S_\sigma, TU(\sigma(T'')), u[\alpha \leftarrow \sigma(r_1)]) \approx v, T \\
&= && TU(u/\alpha \approx \sigma(l_1[\beta \leftarrow r_2])), S_\sigma, u[\alpha \leftarrow \sigma(r_1)]) \approx v, T \stackrel{*}{\Longrightarrow} S_{\sigma'} \cup \widehat{S}.
\end{aligned}
$$

In other respects the proof is as above. This completes the induction and the proof of the lemma. $\square$

Readers interested in further discussion of this method should consult the paper [39] and the thesis [81].

6.6 Previous Work

In this section we compare our approach to E-unification to current and past research in this field. As remarked in Chapter §4, Since the work of Plotkin [131], most of the energy of researchers in this field has been directed either toward (i) isolating and investigating the E-unification problem in specific theories such as commutativity, associativity, etc., and various combinations of such specific axioms, and (ii) investigating the E-unification problem in the presence of canonical rewrite systems. There has been some work as well on various extensions to the latter.

The first area of research has not concerned us in this monograph, since we have been interested only in more general forms of E-unification. The second area represents the most general form of E-unification which has been thoroughly investigated to date (but see also [66]).

Narrowing and its refinements represent a very clean and elegant solution to an important subclass of E-unification problems, and we do not claim to have improved upon these results. Instead we view our research as an attempt to place these results in a more general context, by showing in a very abstract way how the same proof techniques used in narrowing may be applied to our more general problem. We should in particular note that Martelli, Moiso, and Rossi have presented an E-unification procedure using a set of transformations much the same as our set $\mathcal{T}$, but they attempted to prove completeness only in the context of canonical systems.

The work of Kirchner [91] attempts to extend the basic paradigm of E-unification in canonical theories by adapting the approach of Martelli and Montanari [109] to standard unification which uses the operations of merging and decomposition over multiequations to find mgu's in ordered form; by respecting the ordering of variable dependencies among the various terms, one may avoid explicit application of substitutions, and so Variable Elimination is not used. Kirchner expands this basic method by defining conditions under which decomposition may be done in the presence of equations, and by defining a new operation on multiequations, called mutation, which is dependent upon the theory under consideration. He extends the procedure for canonical theories by showing that if a theory permits the use of variable dependency orderings to avoid explicit substitution (such a theory is termed *strict*), and if a mutation operation can be deduced, then his procedure returns a complete set of E-unifiers. He then gives a general strategy for deriving the mutation operation via a critical pair computation, and hence a way of automating the creation of specialized E-unification procedures. As an example this strategy is applied to the class of *syntactic* theories, which basically allow complete sets of E-unifiers to be found by allowing at most one rewrite at the root between any two terms. Our approach to E-unification owes much to Kirchner's initial inspiration to adapt the method of transformations to E-unification, but our motivations are very different. We have used only the abstract notion of transformations on equation systems, and not the technique of multiequations. Our research concerns not the derivation of specific procedures, but the abstract analysis of the general case. It is not surprising, then, that we can subsume the methods of Kirchner in an abstract way. We could optimize our procedure for syntactic theories, for example, by simply allowing at most one root rewrite between any two terms. As in the case of narrowing, however, our general procedure is not likely to be as suitable for specific theories as specially designed procedures, although in an absolute sense it subsumes them.

Another form of more general E-unification which has been investigated using the method of transformations, for example by Holldobler [66], is the problem of E-unification in the presence of a confluent set of rewrite rules. Unfortunately, narrowing is incomplete for this class of theories, as can be seen by considering the rewrite system $R = \{a \rightarrow fa\}$ and the system $fx \approx ffx$, both over the signature $\Sigma = \{f, a\}$. Then clearly for any s and t, $s \xleftrightarrow{*}_R t$ iff either $s = t$ or $s = f^n a$ and $t = f^m a$ for some $n \neq m$, and thus in the second case if $n < m$ we have a sequence $s \xrightarrow{m-n}_R t$, and similarly in the other direction if $m < n$. Thus R is confluent (note that R is not simply ground confluent). But there is no possible narrowing step out of either fx or ffx, although any σ of the form $[f^k a/x]$ for any $k \geq 0$ is an R-unifier. As has been noted, for example in [121] and [67], for confluent but non-terminating systems, narrowing is complete over the set of solutions in normal form (if such exist—in the previous example there are none). Thus it seems that ground confluence with respect to some reduction ordering $\succ$ is needed if the transformations are to be applied oriented, as they are in Holldober's paper.

In general, our approach to E-unification, although heavily indebted to many researchers in this field, is fundamentally different. Whereas the previous work in this field has concentrated on elucidating the structure of specific E-unification problems or in gradually expanding the class of theories for which complete E-unification procedures exist, our research has concentrated on finding a very general method for which a rigorous completeness proof was available, and then attempting to find techniques to prove the completeness of restricted versions of this method.

6.7 Eager Variable Elimination

We discuss in this section an interesting open problem which remains in our research on general E-unification. Notice that in our general discussion of E-unification in Chapter §5, we prove the completeness of the method via a strategy which applies transformation (C) *only* to trivial proofs $(x = t)$ in which no rewrite steps occur. If the proof $(x * t)$ contains rewrite steps, we use transformation (D) or (E). This corresponds in the transformations on systems to non-deterministically allowing an equation $x \approx t$ where $x \notin Var(t)$ to be transformed by either Variable Elimination, Root Rewriting, or Root Imitation in the set $\mathcal{BT}$ or, alternately, by either Variable Elimination or Lazy Paramodulation in the set $\mathcal{T}$. The strategy of *Eager Variable Elimination* is to always apply Variable Elimination to an equation (if possible) instead of Root Rewriting or Root Imitation (or Lazy Paramodulation in the case of $\mathcal{T}$). In other words, we never look for rewrites below the root of an equation $x \approx t$ if $x \notin Var(t)$, and can immediately eliminate x via Variable Elimination. The question of whether such a set of transformations is complete is still open.

In fact, our original formulation of E-unification via transformations used this strategy, but a difficulty arose in finding a measure on which to base our completeness proof. The problem is that—no matter what formalism is used for E-unification proofs—performing Variable Elimination on an equation which needs rewrite steps between $\theta(x)$ and $\theta(t)$ will have to incorporate these steps into the proof wherever x is replaced by t. The effect is that the same equation may end up being duplicated many times. Then, if variables are renamed in duplicated equations to avoid clashes, potentially not only the number of rewrite steps in the new system is increased, but also the number of unsolved variables; but if duplicate equations are not renamed, it must be ensured that no variable clashes will ever occur in any later sequence of transformations.

Actually, the notion of an equational proof tree was developed to clarify these issues, but we were not able to prove the correctness or termination of this new set of proof transformations, and so were led to the approach presented above in Chapter §6 find useful restrictions on our transformations.

The literature has mostly overlooked this problem, and, as it is deceptively simple at first glance, it is generally assumed to be true. Martelli et al. [111] claim the completeness of such a strategy in the context of canonical rewrite systems. However, because their proof lacks many details, including a measure for a rigorous induction, we are unable to check the validity of their argument about Variable Elimination. Holldobler [66] claims the completeness of a set of transformations equivalent to our system BT with Eager Variable Elimination. As remarked above, his proof contains a gap, and no rigorous analysis of Variable Elimination is presented. Using the techniques developed in this chapter, we believe that Holldobler's completeness proof can be partially patched, but we do not believe that the transformations are complete if Eager Variable Elimination is performed. We should remark that Kirchner has avoided this whole problem by examining only those theories in which Variable Elimination can be avoided by the use of variable dependency orderings.

6.8 Current and Future Work

The work discussed in this chapter has been extended in various ways since the original paper [54]. Various other researchers have given inference systems for the general E-unification problem in addition to Dougherty and Johann (discussed in Section §6.5). The system of Jouannaud and Hsiang, first presented at the annual Unification Workshop in 1988, and given in the survey [83], is interesting in that it breaks up the inference rules into more cases (for example, various failure rules are given), but a completeness proof is still forthcoming. An interesting preliminary report on a method for improving the lazy paramodulation method was given by Bertrand Delsart at the Unification Workshop in Barbizon in 1991; the basic idea here is to use a limited amount of forward reasoning (i.e., completion of the set of equations) in order to restrict the use of the lazy paramodulation rule. Another general E-unification inference system which uses forward reasoning is studied in [144].

The issue of backward vs. forward reasoning (represented in this monograph by the results in Sections §6.2 – §6.5 vs. Sections §6.3 – §6.4) is an interesting one which has been studied for a long time in the context of resolution systems without equality. Backward reasoning consists of only allowing inferences which are in some way derived from the goal. In equational logic, forward reasoning is best exemplified by completion, where there is no reference to any goal while critical pairs are being generated; and backward reasoning is used for example when checking for equality between two terms wrt a canonical rewriting system. In resolution theorem provers [139], the strategy of backward reasoning is best exemplified by the *set of support* strategy [157]. Formally, this strategy seeks a refutation for an unsatisfiable set of clauses S by isolating a set $T \subseteq S$ such that $S - T$ is satisfiable (for example, $S - T$ might be the set of hypotheses, and T the negation of the theorem to be proved) and forbidding inferences which take place only among the clauses in $S - T$ (for a good discussion of the crucial importance of this, see [130]).

Unfortunately, this restriction can not be used in the presence of equational axioms if paramodulation into variables is not allowed. For example,

$$\{f(a, b) \doteq a, \ a \doteq b\} \models \exists x.\, f(x, x) \doteq x,$$

so that the set of clauses

$$\{ \ \{f(a, b) \doteq a\}, \ \{a \doteq b\}, \ \{\neg f(x, x) \doteq x\} \ \}$$

is unsatisfiable, but if we pick the obvious set of support, namely the third clause, then we can only obtain a refutation by paramodulating into a variable.[10] This is the motivation for the goal-directed transformations on systems approach taken in this book, and recently the author and Christopher Lynch have extended the principles laid out here to both Horn-Clause logic [149] and to full first-order clausal logic with equality [151]. The basic approach is similar to the proof given here; in particular, in both papers, we need a lemma similar to Lemma 6.2.7 which shows how to convert a proof which is in some ways "canonical" to a proof which uses lazy paramodulation (in fact we work within the context of Relaxed Paramodulation in both these papers). One interesting feature of the last mentioned paper is that in order to prove completeness, it was necessary to extend the basic strategy for narrowing (such as we have used above) to the full paramodulation calculus; the resulting inference system, naturally called Basic Paramodulation, is discussed in [152].

---

[10] Technically, of course, we could define the set of support as the whole set, but this is uninteresting, since then there is no restriction.

6.9 Conclusion

Although research in E-unification has grown tremendously in the past 15 years, for some reason the problem of general E-unification in arbitrary theories has been neglected. This is unfortunate, since progress in any area of science is often frustrated when fundamental issues of the basic paradigm are not well understood. In this section of the monograph we attempted to provide a rigorous paradigm for the study of complete procedures for general E-unification by adopting the method of transformations on systems of terms and showing how a basic set $\mathcal{BT}$ of very general transformations for E-unification corresponds to certain transformations on equational proof trees. In this context, the completeness of our method is easily shown, and highlights a number of features, such as the problem with eager variable elimination discussed above, which are not obvious in completeness proofs using other techniques. In order to make this method efficient enough to be implemented, we then showed how restrictions may be placed on this basic set to obtain a set $\mathcal{T}$, thereby increasing its efficiency while retaining completeness for arbitrary equational theories. The method of proof here was adapted from unfailing completion, and showed that we need not ever rewrite at variable occurrences, which not only eliminates the guessing of functional reflexivity axioms and the potential for infinite recursion on Root Imitation, but also prunes out a large number of useless rewrite sequences. In addition, we showed how other more general forms of E-unification, such as narrowing, can be simulated by our method, by demonstrating that the set of $\mathcal{T}$-transformations is complete for a set R of ground Church-Rosser rewrite rules, and also that the strategy of surreduction plus the simulation of critical pair computation is complete.

In the next chapter, we explore another general paradigm for unification, this time for higher-order terms, again using the method of transformations on systems of terms.

CHAPTER 7

HIGHER ORDER UNIFICATION

Higher-order unification is a method for unifying terms in the Simple Theory of Types [28], that is, given two typed lambda-terms e_1 and e_2, finding a substitution σ for the free variables of the two terms such that $\sigma(e_1)$ and $\sigma(e_2)$ are equivalent under the conversion rules of the calculus. As discussed above, this problem is fundamental to theorem proving in higher-order contexts (see Chapter §1 for references). In this chapter, we adapt the method of transformations to higher-order unification.

7.1 Preliminaries

In order that this section of the monograph be self-contained, we present here a number of basic definitions and results related to the typed lambda calculus, including a detailed treatment of the notion of a substitution. Our notation and approach is basically consistent with [16], [49], [65], and [72].

Definition 7.1.1 Given a set T_0 of basic types (e.g., such as *int*, *bool*, etc.) we define the set of types T inductively as the smallest set containing T_0 and such that if $\alpha, \beta \in T$, then $(\alpha \to \beta) \in T$.

The type $(\alpha \to \beta)$ is that of a function from objects of type α to objects of type β. We assume that the type constructor $\to$ associates to the right, and we shall often write type expressions such as $(\alpha_1 \to (\alpha_2 \to \ldots (\alpha_n \to \beta) \ldots))$ in the form $\alpha_1, \ldots, \alpha_n \to \beta$, with β an arbitrary type.

Definition 7.1.2 Let us assume given a set Σ of symbols, which we call *function constants*, each symbol f having a unique type $\tau(f)$ from T. For each type $\alpha \in T$, we assume given a countably infinite set of variables of that type, denoted V_α, and let $V = \bigcup_{\tau \in T} V_\tau$. Furthermore, let the set of *atoms* A be defined as $V \cup \Sigma$. The set $\mathcal{L}$ of lambda-terms is inductively defined as the smallest set containing A and closed under the rules of function application and lambda-abstraction, namely,

(i) If $e_1 \in \mathcal{L}$ has type $\alpha \to \beta$, and $e_2 \in \mathcal{L}$ has type α, then $(e_1 e_2)$ is a member of $\mathcal{L}$ of type β.

(ii) If $e \in \mathcal{L}$ has type β and $x \in V_\alpha$ then $(\lambda x.e)$ is a member of $\mathcal{L}$ of type $\alpha \to \beta$.

We shall denote the type of a term e by $\tau(e)$.

By convention, application associates to the left, with the result that a term $(\ldots((e_1 e_2)e_3)\ldots e_n)$ may be represented as $(e_1 e_2 \ldots e_n)$. In general we represent a sequence of lambda abstractions

$$\lambda x_1.(\lambda x_2.(\ldots(\lambda x_n.e)\ldots))$$

in the form $\lambda x_1 \ldots x_n.e$, where e is either an application or an atom. We shall often drop superfluous parentheses when there is no loss of clarity, and will use square brackets if necessary; also we follow the convention that the dot includes as much right context as possible in the scope of its binder, so that, e.g., a term $\lambda x.stu$ is to be interpreted as $(\lambda x.((st)u))$.

Definition 7.1.3 In a term $\lambda x_1 \ldots x_n.e$ where e is either an application or an atom, we call e the *matrix* of the term, the object $\lambda x_1 \ldots x_n$ is the *binder* of the term, and the occurrences of the variables are called *binding occurrences* of these variables. We define the *size* of a term u, denoted $|u|$, as the number of atomic subterms of u. A variable x occurs *bound* in a term e if e contains some subterm of the form $\lambda x.e'$, in which case the term e' is called the *scope* of this binding occurrence of x. A variable x occurs *free* in e if it is a subterm of e but does not occur in the scope of a binding occurrence of x. The set of free variables of a term e is denoted by $FV(e)$.

Definition 7.1.4 The order of a term or a variable is just the order of its type, where the *order* of a type φ is defined as

$$Ord(\varphi) = \begin{cases} 1, & \text{if } \varphi \in \mathcal{T}_0; \\ max(Ord(\alpha)+1, Ord(\beta)), & \text{if } \varphi = \alpha \to \beta. \end{cases}$$

A *language of order n* is one which allows terms of order at most n.

This formalizes the usual convention that a first-order term denotes an individual, a term of second order denotes a function on individuals, etc.

Convention: In what follows we denote types by α, β, γ, and φ; constants of primitive type by b and c; constants of functional type by f, g, and h; variables of arbitrary type by x, y, and z, and arbitrary atoms by a. We shall often represent free variables of functional type by the letters F, G,

H, and Y. Lambda terms will be denoted by e, r, s, t, u, v, and w. We shall, in the interest of clarity, omit type information whenever possible, since it is inferrable from context in the cases we consider.

The 'computation rules' of the lambda calculus are as follows.

Definition 7.1.5 Let $u[t/x]$ denote the result of replacing each free occurrence of x in u by t, and $BV(t)$ be the set of bound variables in t. We have three rules of *lambda conversion*.

(i) (α-conversion) If $y \notin FV(t) \cup BV(t)$, then

$$(\lambda x.\, t) \;\succ_\alpha\; (\lambda y.\, (t[y/x])).$$

(ii) (β-conversion)
$$((\lambda x.\, s)\, t) \;\succ_\beta\; s[t/x].$$

(iii) (η-conversion)[1] If $x \notin FV(t)$, then

$$(\lambda x.\, (t\, x)) \;\succ_\eta\; t.$$

The term on the left side of each of these rules is called a *redex*. A term t which contains no β-redices is called a β-*normal form*, and η-normal forms and $\beta\eta$-normal forms are defined similarly. If we denote by $e[s]$ a lambda term with some distinguished occurrence of a subterm s, then let $e[t]$ denote the result of replacing this single subterm by the term t, where $\tau(s) = \tau(t)$. We define the relation $\longrightarrow_\alpha$ as

$$e[s] \longrightarrow_\alpha e[t] \quad \text{iff} \quad s \succ_\alpha t,$$

and similarly for $\longrightarrow_\beta$ and $\longrightarrow_\eta$. We define $\longrightarrow_{\beta\eta}$ as $\longrightarrow_\beta \cup \longrightarrow_\eta$. We also define the symmetric closure $\longleftrightarrow$, the transitive closure $\xrightarrow{+}$, and the symmetric, reflexive, and transitive closure $\xleftrightarrow{*}$ of each of these relations in the obvious fashion. The relations $\xleftrightarrow{*}_\beta$, $\xleftrightarrow{*}_\eta$, and $\xleftrightarrow{*}_{\beta\eta}$ are called β-, η-, and $\beta\eta$-*equivalence* respectively.

It is easy to show that the type of a lambda term is preserved under these rules of lambda conversion.

Definition 7.1.6 We say that s *is substitutible for* x *in* t if, for every subformula $\lambda y.\, t'$ of t, if $y \in FV(s)$ then $x \notin FV(t')$.

---

[1] This rule is a special case of the the axiom of extensionality, viz., $\forall f, g(\forall x(f(x) = g(x)) \implies f = g)$, which asserts that two functions are equal if they behave the same on all arguments, regardless of their syntactic representation.

The motivation for this notion is that no free variable capture will take place if s is substituted for x in t'. (The problem with this *free variable capture* is that it violates the fundamental meaning of scope and the binding of variables; in [16], for the untyped calculus it is shown that if this is allowed, the calculus becomes inconsistent in the sense that any two terms are equivalent.) In the β-conversion rule, in the pathological case that s is not substitutible for x in t, i.e., x occurs in t in the scope of some binding occurrence of a variable which is free in s, then there is always a sequence $(\lambda x.t)\, s \xrightarrow{+}_\alpha (\lambda x.t')\, s \longrightarrow_\beta t'[s/x]$, where s is substitutible for x in t'. Thus, for simplicity and without loss of generality we adopt the following assumption.

Convention: We assume in the following that in the set of terms being discussed, the set of all free variables is distinct from the set of all bound variables. (This allows us to be 'naive' in our use of β-conversion and substitution; for another approach, see [65].) In fact, in the rest of this chapter, all comparisons of lambda terms are modulo α-conversion, which will allow us to represent lambda binders using 'generic' variables $x_1, \ldots, x_k$ unless confusion would result. By abuse of notation, using this naive approach and following our representation of a sequence of lambda abstractions as a term $\lambda x_1 \ldots x_k.\, u$, we shall consider the conversion of redices involving such terms as a *single* reduction step instead of k steps, e.g.,

$$(\lambda x_1 \ldots x_k.\, u)\, v_1 \ldots v_k \longrightarrow_\beta u[v_1/x_1, \ldots, v_k/x_k]$$

instead of $(\lambda x_1 \ldots x_k.\, u)\, v_1 \ldots v_k \xrightarrow{k}_\beta u[v_1/x_1, \ldots, v_k/x_k]$.

Definition 7.1.7 The calculus which admits only the β-rule as a computation rule we call the *typed β-calculus* and the calculus which also admits the η-rule is called the *typed $\beta\eta$-calculus*.

In this chapter, we wish to give an abstract method for higher-order unification which presents the fundamental logical issues as clearly as possible, and for this purpose we feel it is sufficient to develop the notion of unification of terms in the typed $\beta\eta$-calculus. This is a natural assumption in practice, and all higher-order theorem proving systems known to the author use this weak form of extensionality. The reader interested in the details of the non-extensional case may consult [72].

Two of the major results concerning this calculus are the following.

Theorem 7.1.8 (Strong Normalization) Every sequence of $\beta\eta$-reductions is finite.

Theorem 7.1.9 (Church-Rosser Theorem) If $s \xleftrightarrow{*}_{\beta\eta} t$ for two lambda terms s and t, then there must exist a term u such that $s \xrightarrow{*}_{\beta\eta} u \xleftarrow{*}_{\beta\eta} t$.

(Proofs of these may be found in [65].) Each of these theorems remains true when restricted to just η-conversion or just β-conversion. One of the important consequences of these two results is that for each term t there exists a unique (up to α-conversion) term t' such that $t \xrightarrow{*}_{\beta\eta} t'$ with t' in $\beta\eta$-normal form, and similarly for the restriction to just β- or just η-reduction. Another consequence is that the β-, η-, or $\beta\eta$-equivalence of two arbitrary terms may be decided by checking if the corresponding normal forms of the two terms are equal. For example, if we denote the unique β-normal form of a term t by $t{\downarrow}$, then $s \xleftrightarrow{*}_{\beta} t$ iff $s{\downarrow} = t{\downarrow}$.

Convention: We shall in general assume that terms under discussion are in β-normal form unless otherwise stated. In particular, each term in β-normal form may be represented in the form $\lambda x_1 \ldots x_n (a\, e_1 \ldots e_m)$, where the *head* a is an atom, i.e., a is either a function constant, bound variable, or some variable free in this term, and the terms $e_1, \ldots, e_m$ are in the same form. By analogy with first-order notation, such a term will be denoted $\lambda x_1 \ldots x_n . a(e_1, \ldots, e_m)$. As an abbreviation, we represent lambda terms using something like a 'vector' notation for lists, so that $\lambda x_1 \ldots x_n . e$ will be represented by $\lambda \overline{x_n} . e$. Furthermore, this principle will be extended to lists of terms, so that $\lambda \overline{x_n} . f(e_1, \ldots, e_m)$ will be represented as $\lambda \overline{x_n} . f(\overline{e_m})$, and we shall even sometimes represent a term such as

$$\lambda \overline{x_k} . a(y_1(\overline{x_k}), \ldots, y_n(\overline{x_k}))$$

in the form $\lambda \overline{x_k} . a(\overline{y_n(\overline{x_k})})$.

Definition 7.1.10 A term whose head is a function constant or a bound variable is called a *rigid* term; if the head is a free variable it will be called a *flexible* term. (For example, the term $\lambda x . F(\lambda y . y(x, a), c)$ is flexible, but both of its immediate subterms are rigid.)

As remarked above, we consider in this chapter only the problem of unifying terms in the $\beta\eta$-calculus, and since our analysis proceeds by examining the manner in which substitution and subsequent β-reduction makes two terms identical, we need not explicitly consider the role of η-reduction. The formal justification for this is given by the following result.

Lemma 7.1.11 For any two terms s and t, we have $s \xrightarrow{*}_{\beta\eta} t$ iff there exists a term u such that $s \xrightarrow{*}_{\beta} u \xrightarrow{*}_{\eta} t$.

(For a proof see [16].) As a consequence, we can decide $\beta\eta$-equivalence by reducing terms to their β-normal forms, and then testing for η-equivalence, that is, $s \xleftrightarrow{*}_{\beta\eta} t$ iff $s{\downarrow} \xleftrightarrow{*}_{\eta} t{\downarrow}$. This allows us to 'factor out' η-conversion, by considering only η-equivalence classes of terms. We shall use the following means of representing such classes by canonical representatives (due to [72]).

Definition 7.1.12 Let $e = \lambda x_1 \ldots x_n . a(e_1, \ldots, e_m)$ be a term in β-normal form of type $\alpha_1, \ldots, \alpha_n, \alpha_{n+1}, \ldots, \alpha_{n+k} \to \beta$, with $\beta \in T_0$. The η-expanded form of e, denoted by $\eta[e]$, is produced by adding k new variables of the appropriate types to the binder and the matrix of the term, and (recursively) applying the same expansion to the subterms, to obtain

$$\lambda x_1 \ldots x_n x_{n+1} \ldots x_{n+k} . a(\eta[e_1], \ldots, \eta[e_m], \eta[x_{n+1}], \ldots, \eta[x_{n+k}]),$$

where $\tau(x_{n+i}) = \alpha_{n+i}$ for $1 \le i \le k$.

This is effectively the normal form of a term under the converse of the η-reduction rule (so that $\eta[e] \xrightarrow{*}_{\eta} e$) and is *only* defined on a term already in β-normal form. It is easy to show that in an η-expanded form, every atom appears applied to as many arguments as allowed by its type, and that the matrices of all subterms are of base types. This form is more useful than the η-normal form because it makes the type of the term and all its subterms more explicit, and is therefore a convenient syntactic convention for representing the congruence class of all terms equal modulo the $\eta-$rule. It is easy to show, by structural induction on terms, that these expanded forms always exist and are unique (up to α-conversion), so that for any two terms s and t in β-normal form, we have $s \xrightarrow{*}_{\eta} t$ iff $\eta[s] = \eta[t]$ (see [72], Lemma 4.3). Thus, we have a Church-Rosser theorem in the following form.

Theorem 7.1.13 For every two terms s and t, we have $s \xleftrightarrow{*}_{\beta\eta} t$ iff $\eta[s{\downarrow}] = \eta[t{\downarrow}]$.

Definition 7.1.14 Let $\mathcal{L}_{exp}$ be defined as the set of all η-expanded forms, i.e., $\mathcal{L}_{exp} = \{\eta[e{\downarrow}] \mid e \in \mathcal{L}\}$. Define the set $\mathcal{L}_\eta$ as the smallest subset of $\mathcal{L}$ containing $\mathcal{L}_{exp}$ and closed under application and lambda abstraction, i.e., $(e_1 e_2)$ and $\lambda x . e_1$ are in $\mathcal{L}_\eta$ whenever $e_1 \in \mathcal{L}_\eta$ and $e_2 \in \mathcal{L}_\eta$.

The essential features of $\mathcal{L}_{exp}$ and $\mathcal{L}_\eta$ which will allow us to restrict our attention to η-expanded forms are proved in the next lemma, which is from [72].

Lemma 7.1.15 For every variable x and every pair of terms e and e' of the appropriate types:

(1) $e, e' \in \mathcal{L}_{exp}$ implies that $(\lambda x . e) \in \mathcal{L}_{exp}$ and $(ee')\!\downarrow\in \mathcal{L}_{exp}$;

(2) $e \in \mathcal{L}_\eta$ implies that $e\!\downarrow\in \mathcal{L}_{exp}$;

(3) $e, e' \in \mathcal{L}_\eta$ implies that $(\lambda x . e) \in \mathcal{L}_\eta$ and $(ee') \in \mathcal{L}_\eta$;

(4) $e \in \mathcal{L}_\eta$ and $e \overset{*}{\longrightarrow}_\beta e'$ implies that $e' \in \mathcal{L}_\eta$;

(5) $e, e' \in \mathcal{L}_\eta$ implies that $e'[e/x] \in \mathcal{L}_\eta$.

These closure conditions for $\mathcal{L}_\eta$ (not all of which are satisfied by the set of η-normal forms) formally justify our leaving the η-rule implicit in the following sections by developing our method for higher-order unification in the language $\mathcal{L}_\eta$ and considering explicitly only β-conversion as a computation rule.[2] The reader interested in a more detailed treatment of these matters, including proofs of the previous results, is referred to [72] for details.

We now formalize the general notion of substitution of lambda terms for free variables in the $\beta\eta$-calculus, after which we show how this may be specialized to substitutions over $\mathcal{L}_{exp}$.

Definition 7.1.16 A *substitution* is any (total) function $\sigma : V \rightarrow \mathcal{L}$ such that $\sigma(x) \neq x$ for only finitely many $x \in V$ and for every $x \in V$ we have $\tau(\sigma(x)) = \tau(x)$. Given a substitution σ, the *support* (or *domain*) of σ is the set of variables $D(\sigma) = \{x \mid \sigma(x) \neq x\}$. A substitution whose support is empty is termed the *identity substitution*, and is denoted by Id. The set of variables *introduced by* σ is $I(\sigma) = \bigcup_{x \in D(\sigma)} FV(\sigma(x))$.

A subtle point of this definition is that substitutions are total functions which are non-trivial over only a finite number of variables; over the rest of V they simply map variables to themselves. Given a substitution σ, if its support is the set $\{x_1, \ldots, x_n\}$, and if $t_i = \sigma(x_i)$ for $1 \leq i \leq n$, then σ is also denoted by listing its bindings explicitly: $[t_1/x_1, \ldots, t_n/x_n]$. Given a term u, we may also denote $\sigma(u)$ as $u[t_1/x_1, \ldots, t_n/x_n]$.

Definition 7.1.17 A substitution ρ is a *renaming substitution away from* W if $\rho(x)$ is a variable (modulo η-conversion) for every $x \in D(\rho)$, $I(\rho) \cap W = \emptyset$, and for every x and y in $D(\rho)$, $\rho(x) \overset{*}{\longleftrightarrow}_\eta \rho(y)$ implies that $x = y$. If W is unimportant, then ρ is simply called a *renaming*. The *restriction*

---

[2] In fact, we shall depart from our convention in the interests of simplicity only when representing terms which are (up to η-conversion) variables, e.g., $\lambda xy . F(x, y)$. In some contexts, such as solved form systems, we wish to emphasize their character as variables, and will represent them as such, e.g., just F. In these cases, we shall be careful to say that 'F is (up to η-conversion) a variable,' etc.

of a substitution σ to some W', denoted $\sigma|_{W'}$, is the substitution σ' such that

$$\sigma'(x) = \begin{cases} \sigma(x), & \text{if } x \in W'; \\ x, & \text{otherwise.} \end{cases}$$

Since $\mathcal{L}$ is freely generated, every substitution $\sigma : V \to \mathcal{L}$ has a unique extension $\hat{\sigma} : \mathcal{L} \to \mathcal{L}$ defined recursively as follows.

Definition 7.1.18 Let σ_{-x} denote the substitution $\sigma|_{D(\sigma)-\{x\}}$. For any substitution σ,

$$\hat{\sigma}(x) = \sigma(x) \text{ for } x \in V;$$
$$\hat{\sigma}(a) = a \text{ for } a \in \Sigma;$$
$$\hat{\sigma}(\lambda x.\, e) = \lambda x.\, \widehat{\sigma_{-x}}(e);$$
$$\hat{\sigma}((e_1\, e_2)) = (\hat{\sigma}(e_1)\, \hat{\sigma}(e_2)).$$

Thus a substitution has an effect only on the *free* variables of a term. In the sequel, we shall identify σ and its extension $\hat{\sigma}$. Note that by our assumption that the sets of bound variables and free variables in any context are disjoint, no variable capture will ever take place by application of a substitution. It is easy to show that the type of a term is unchanged by application of an arbitrary substitution.

Remark: It is important to note that by $\sigma(e)$ we denote the result of applying the substitution σ to e *without* β-reducing the result; we shall denote by $\sigma(e)\!\downarrow$ the result of applying the substitution and then reducing the result to β-normal form. This rather non-standard separation we impose between substitution and the subsequent β-reduction is useful because we wish to examine closely the exact effect of substitution and β-reduction on lambda terms in a later section.

Definition 7.1.19 The *union* of two substitutions σ and θ, denoted by $\sigma \cup \theta$, is defined by

$$\sigma \cup \theta(x) = \begin{cases} \sigma(x), & \text{if } x \in D(\sigma); \\ \theta(x), & \text{if } x \in D(\theta); \\ x, & \text{otherwise,} \end{cases}$$

and is only defined if $D(\sigma) \cap D(\theta) = \emptyset$. The *composition* of σ and θ is the substitution denoted by $\sigma \circ \theta$ such that for every variable x we have $\sigma \circ \theta(x) = \hat{\theta}(\sigma(x))$. Note carefully that we denote composition from *left to right*.

Definition 7.1.20 Given a set W of variables, we say that two substitutions σ and θ are *equal over* W, denoted $\sigma = \theta[W]$, iff $\forall x \in W$, $\sigma(x) = \theta(x)$. Two substitutions σ and θ are *β-equal over* W, denoted $\sigma =_\beta \theta[W]$ iff $\forall x \in W$, $\sigma(x) \overset{*}{\longleftrightarrow}_\beta \theta(x)$, or, equivalently, $\sigma(x){\downarrow} = \theta(x){\downarrow}$. The relations $=_\eta$ and $=_{\beta\eta}$ are defined in the same way but using $\overset{*}{\longleftrightarrow}_\eta$ and $\overset{*}{\longleftrightarrow}_{\beta\eta}$. We say that σ is *more general than* θ *over* W, denoted by $\sigma \leq \theta[W]$, iff there exists a substitution η such that $\theta = \sigma \circ \eta[W]$, and we have $\sigma \leq_\beta \theta[W]$ iff there exists some η' such that $\theta =_\beta \sigma \circ \eta'[W]$, and $\leq_\eta$ and $\leq_{\beta\eta}$ are defined analogously. When W is the set of all variables, we drop the notation $[W]$. If neither $\sigma \leq_{\beta\eta} \theta$ nor $\theta \leq_{\beta\eta} \sigma$ then σ and θ are said to be *independent*.

The comparison of substitutions modulo β-, η-, and $\beta\eta$-conversion is formally justified by the following lemma, which is easily proved by structural induction on terms:

Lemma 7.1.21 If σ and θ are arbitrary substitutions such that either $\sigma =_\beta \theta$, $\sigma =_\eta \theta$, or $\sigma =_{\beta\eta} \theta$, then for any term u we have either $\sigma(u) \overset{*}{\longleftrightarrow}_\beta \theta(u)$, $\sigma(u) \overset{*}{\longleftrightarrow}_\eta \theta(u)$, or $\sigma(u) \overset{*}{\longleftrightarrow}_{\beta\eta} \theta(u)$, respectively.

We now show that we can develop the notion of substitution wholly within the context of the language $\mathcal{L}_\eta$ developed above without loss of generality.

Definition 7.1.22 A substitution θ is said to be *normalized* if $\theta(x) \in \mathcal{L}_{exp}$ for every variable $x \in D(\theta)$.

We can assume without loss of generality that no normalized substitution has a binding of the form $\eta[x]/x$ for some variable x. A normalized renaming substitution has the form $[\eta[y_1]/x_1, \ldots, \eta[y_n]/x_n]$; the effect of applying such a substitution and then β-reducing is to rename the variables $x_1, \ldots, x_n$ to $y_1, \ldots, y_n$. The justification for using normalized substitutions is given by the following corollary of Lemma 7.1.15.

Corollary 7.1.23 If θ is a normalized substitution and $e \in \mathcal{L}_{exp}$, then $\theta(e) \in \mathcal{L}_\eta$ and $\theta(e){\downarrow} \in \mathcal{L}_{exp}$.

It is easy to show that if σ and θ are normalized, then $\sigma =_{\beta\eta} \theta$ iff $\sigma = \theta$ and if θ' is the result of normalizing θ, then $\theta' =_{\beta\eta} \theta$.

Convention: In general, substitutions are assumed to be normalized in the rest of this chapter, allowing us to factor out η-equivalence in comparing substitutions, so that we may, e.g., use $\leq_\beta$ instead of $\leq_{\beta\eta}$. In fact, the composition of two normalized substitutions could be considered to be a normalized substitution as well, so that $\sigma \leq_\beta \theta$ iff $\sigma \leq \theta$, but this need *not* be assumed in what follows. For example, the composition $[\lambda x.\, G(a)/F] \circ [\lambda y.\, y/G]$ is defined as $[\lambda x.\, ((\lambda y.\, y)a)/F, \lambda y.\, y/G]$, *not* as $[\lambda x.\, a/F, \lambda y.\, y/G]$. We shall continue to use $=_\beta$ and $\leq_\beta$ to compare normalized substitutions, although strictly speaking the subscript could be omitted if no composition is involved.

Definition 7.1.24 A substitution σ is *idempotent* if $\sigma \circ \sigma =_{\beta\eta} \sigma$.

A sufficient condition for idempotency is given by[3]

Lemma 7.1.25 A substitution σ is idempotent if $I(\sigma) \cap D(\sigma) = \emptyset$.

That in most contexts we may restrict our attention to idempotent substitutions without loss of generality is demonstrated by our next result, which shows that any substitution is equivalent (over an arbitrarily chosen set of variables) up to renaming to an idempotent substitution.

Lemma 7.1.26 For any substitution σ and set of variables W containing $D(\sigma)$, there exists an idempotent substitution σ' such that $D(\sigma) = D(\sigma')$, $\sigma \leq_{\beta\eta} \sigma'$, and $\sigma' \leq_{\beta\eta} \sigma[W]$.

Proof. Let $D(\sigma) \cap I(\sigma) = \{x_1, \ldots, x_n\}$, let $\{y_1, \ldots, y_n\}$ be a set of *new* variables not occurring in W, $\rho_1 = [y_1/x_1, \ldots, y_n/x_n]$, and $\rho_2 = [x_1/y_1, \ldots, x_n/y_n]$. Now let $\sigma' = \sigma \circ \rho_1$, where clearly $\sigma \leq_{\beta\eta} \sigma'$ and $D(\sigma) = D(\sigma')$ as required. Since $\rho_1 \circ \rho_2 =_{\beta\eta} Id[W \cup I(\sigma)]$, then $\sigma =_{\beta\eta} \sigma \circ \rho_1 \circ \rho_2 =_{\beta\eta} \sigma' \circ \rho_2[W]$, and thus $\sigma' \leq_{\beta\eta} \sigma[W]$. Finally, by our previous lemma, σ' must be idempotent, since $D(\sigma') = D(\sigma)$ is disjoint from $I(\sigma') = (I(\sigma) - \{x_1, \ldots, x_n\}) \cup \{y_1, \ldots, y_n\}$. $\square$

In general the assumption of idempotency simplifies matters. We shall provide specific motivations for the use of idempotent unifiers in the appropriate sections.

The net effect of these definitions, conventions, and results is that we can develop our method for unification of terms in the $\beta\eta$-calculus wholly within $\mathcal{L}_\eta$, leaving η-equivalence implicit in the form of the terms under consideration.

---

[3] In the first-order case, this condition is necessary as well, but in our more general situation we have counter-examples such as $\sigma = [\lambda x.\, F(a)/F]$.

7.2 Higher Order Unification via Transformations

Higher-order unification is more complex than first-order unification due to
the presence of variables of functional type, the notion of scope and bound
variables, and the fact that unification is defined in terms of $\beta\eta$-equivalence.
This additional syntactic complexity has several serious consequences. First
of all, the unification of terms of second-order and higher is undecidable
in general [58]. Next, most general unifiers do not exist in general, and
we must again consider complete sets of unifiers. Finally, due to the complexity of the subproblem of unifying two flexible terms, the search space
for a complete unification procedure may be infinitely branching, which
forbids any reasonable implementation. (Note that the first two of these
consequences are shared by E-unification, but the third is not, as shown in
Chapter §6.)

Our analysis of the problem proceeds by examining the exact fashion in
which substitution and β-reduction makes two terms identical from the top-
down (i.e., from the head to the innermost subterms). We develop from this
a set of non-deterministic transformations extending those of the previous
section, and prove their non-deterministic completeness in an analogous
fashion. In Section §7.3, this is restricted to the problem of preunification.

Definition 7.2.1   The notion of pairs and systems of equations carries
over from the first-order case. A substitution θ is a unifier of two lambda
terms e_1 and e_2 iff $\theta(e_1) \xleftrightarrow{*}_{\beta\eta} \theta(e_2)$.[4]   A substitution is a unifier of a
system S if it unifies each pair in S. The set of all unifiers of S is denoted
$U(S)$ and if S consists of a single pair s,t then it is denoted $U(s,t)$.

This definition is more general than we shall need, in fact, since we
shall develop our approach in $\mathcal{L}_\eta$ in order to factor out η-conversion, as
was formally justified in Section §7.1. Thus for two terms $s, t \in \mathcal{L}_\eta$, we
say that a normalized substitution θ is in $U(s,t)$ iff $\theta(s) \xleftrightarrow{*}_{\beta} \theta(t)$, or,
alternately, if $\theta(s){\downarrow} = \theta(t){\downarrow}$.

A pair of terms is solved in a system S if it is in the form $\eta[x], t$, for
some variable x which occurs only once in S; a system is solved if each of its
pairs is solved. Our only departure from the use of η-expanded form is that
we shall represent pairs of the form $\eta[x], t$ as x, t in order to emphasize
their correspondence to bindings t/x in substitutions, as in the first-order
case of the previous section.

---

[4]   This is in the $\beta\eta$-calculus; in the β-calculus the condition would be $\theta(e_1) \xleftrightarrow{*}_{\beta} \theta(e_2)$.

Example 7.2.2 If $u = f(a, g(\lambda x. G(\lambda y. x(b))))$ and $v = F(\lambda x. x(z))$, then $\theta = [\lambda x_2. f(a, g(x_2))/F, \lambda x_3. x_3(z_2)/G, b/z]$ is in $U(u, v)$, since $\theta(u)\!\downarrow = \theta(v)\!\downarrow$:

$$\theta(u) = f(a, g(\lambda x. [(\lambda x_3. x_3(z_2))(\lambda y. x(b))]))$$
$$\longrightarrow_\beta f(a, g(\lambda x. [(\lambda y. x(b))z_2]))$$
$$\longrightarrow_\beta f(a, g(\lambda x. x(b)))$$
$$\longleftarrow_\beta (\lambda x_2. f(a, g(x_2)))(\lambda x. x(b)) = \theta(v).$$

The basic decidability results concerning higher-order unification are as follows.

Definition 7.2.3 For a given set of function constants Σ, the *unification problem* for the language $\mathcal{L}$ generated by Σ is to decide, for any arbitrary terms $e, e' \in \mathcal{L}$, whether the set $U(e, e')$ is non-empty. The nth-*order unification problem* is to decide the unification problem for an arbitrary language of order n.

For example, in Section §3.3 we showed that the first-order unification problem is decidable. Unfortunately, this does not hold for higher-orders.

Theorem 7.2.4 The second-order unification problem is undecidable.

This result was shown by Goldfarb [58] using a reduction from Hilbert's Tenth Problem; previously, Huet [70] showed the undecidability of the third-order unification problem, using a reduction from the Post Correspondence Problem. These results show that there are second-order (and therefore arbitrarily higher-order) languages where unification is undecidable; but in fact there exist *particular* languages of arbitrarily high-order which have a decidable unification problem. Interestingly, Goldfarb's proof requires that the language to which the reduction is made contain at least one 2-place function constant. It has been shown in [45] that the unification problem for second-order monadic languages (i.e., no function constant has more than one argument place) is decidable, which has applications in certain decision problems concerning the lengths of proofs. A different approach to decidability is taken in [159], where decidable cases of the unification problem are found by showing that the search tree for some problems, although infinite, is regular, and that the set of unifiers can be represented by a regular expression. More generally, it has been shown by Statman [153] that the set of all decidable unification problems is polynomial-time decidable.

Besides the undecidability of higher-order unification, another problem is that—as with E-unification—mgu's may no longer exist, a result first shown by [59]. For example, the two terms $F(a)$ and a have the unifiers $[\lambda x.\, a/F]$ and $[\lambda x.\, x/F]$, but there is no unifier more general than both of these. This leads us to extend the notion of a $mgu(S)[W]$ to the higher-order case by considering *complete sets* of unifiers (cf. Definition 4.1.6). Our definition is a generalization of the one found in [72] to equation systems.[5]

Definition 7.2.5 Given a system S and a finite set W of 'protected' variables, a set U of normalized substitutions is a *complete set of unifiers for S away from W* (which we shall abbreviate by $CSU(S)[W]$) iff
  (i) For all $\sigma \in U$, $D(\sigma) \subseteq FV(S)$ and $I(\sigma) \cap (W \cup D(\sigma)) = \emptyset$;
 (ii) $U \subseteq U(S)$;
(iii) For every $\theta \in U(S)$, there exists some $\sigma \in U$ such that $\sigma \leq_\beta \theta[FV(S)]$.
The first condition is called the *purity condition*, the second the *coherence condition*, and the last the *completeness condition*. If S consists of a single pair u, v then we use the abbreviation $CSU(u,v)[W]$. When W is not significant, we drop the notation $[W]$.

That there is no loss of generality in considering only normalized substitutions may be seen by the fact that any substitution is $\beta\eta$-equal to a normalized substitution. By providing a version of Lemma 3.3.11 for this new context, we see that condition (i) is without loss of generality as well.

Lemma 7.2.6 For any system S, substitution θ, and set of protected variables W, if $\theta \in U(S)$ then there exists some normalized substitution σ such that
  (i) $D(\sigma) \subseteq FV(S)$ and $I(\sigma) \cap (W \cup D(\sigma)) = \emptyset$;
 (ii) $\sigma \in U(S)$;
(iii) $\sigma \leq_{\beta\eta} \theta[FV(S)]$ and $\theta \leq_{\beta\eta} \sigma[FV(S)]$.

Proof. If $\sigma = \theta|_{FV(S)}$ satisfies condition (i), then we have our result trivially. Otherwise, if $I(\theta) = \{x_1, \ldots, x_n\}$ then let $\{y_1, \ldots, y_n\}$ be a set of new variables disjoint from the variables in W, $I(\theta)$, and $FV(S)$ such that $\tau(x_i) = \tau(y_i)$ for $1 \leq i \leq n$. Now define the renaming substitutions $\rho_1 = [\eta[y_1]/x_1, \ldots, \eta[y_n]/x_n]$ and $\rho_2 = [\eta[x_1]/y_1, \ldots, \eta[x_n]/y_n]$, let $\sigma' =$

---

[5] The caveat given in footnote 1 of Chapter §4 applies here also, *mutandis mutatis.* Note that our definition is based on our use of $\mathcal{L}_\eta$; in the version for the $\beta\eta$-calculus, condition (iii) would use $\leq_{\beta\eta}$, and substitutions would not have to be normalized. The original Huet definition of a complete set may also be found in [43] in the context of E-unification.

$\theta \circ \rho_1 |_{FV(S)}$, and then let σ be the normalized version of σ'. Clearly σ satisfies (i), and since $\sigma =_{\beta\eta} \theta \circ \rho_1 [FV(S)]$ we have the second part of (iii). Now, because $\rho_1 \circ \rho_2 =_{\beta\eta} Id[FV(S) \cup I(\theta)]$, we must have $\theta =_{\beta\eta} \theta \circ \rho_1 \circ \rho_2 [FV(S) \cup I(\theta)]$. But then by the fact that $\sigma =_{\beta\eta} \theta \circ \rho_1 [FV(S)]$ we have $\theta =_{\beta\eta} \sigma \circ \rho_2 [FV(S)]$, and so $\sigma \leq_{\beta\eta} \theta[FV(S)]$, proving the first part of (iii). To show (ii), observe that for any $u, v \in S$ we have $\theta(u){\downarrow} = \theta(v){\downarrow}$, and for any term t, we have $\sigma'(t) \xleftrightarrow{\ *\ }_{\beta\eta} \sigma(t)$, and so

$$
\begin{aligned}
\sigma(u) \xleftrightarrow{\ *\ }_{\beta\eta} \sigma'(u) &= \rho_1(\theta(u)) \xrightarrow{\ *\ }_{\beta\eta} \rho_1(\theta(u){\downarrow}) \\
&= \rho_1(\theta(v){\downarrow}) \xleftarrow{\ *\ }_{\beta\eta} \rho_1(\theta(v)) \\
&= \sigma'(v) \xleftrightarrow{\ *\ }_{\beta\eta} \sigma(v),
\end{aligned}
$$

which shows that $\sigma \in U(S)$. $\quad\square$

This shows us that for any S and W, the set of all normalized unifiers satisfying condition (i) and (ii) of Definition 7.2.5 is a $CSU(S)[W]$, and so in particular there is no loss of generality in considering only normalized, idempotent unifiers θ such that $D(\theta) \cap I(\theta) = \emptyset$ in what follows. This will simplify our presentation.

Finally, we examine the relevance of solved form systems in $\mathcal{L}_\eta$.

Lemma 7.2.7 If $S = \{x_1, t_1, \ldots, x_n, t_n\}$ is a system in solved form, then $\{\sigma_S\}$ is a $CSU(S)[W]$ for any W such that $W \cap FV(S) = \emptyset$.

Proof. The first two conditions in Definition 7.2.5 are satisfied, since σ_S is an idempotent *mgu* of S, $W \cap FV(S) = \emptyset$, and $I(\sigma_S) \subseteq FV(S)$. Now, if $\theta \in U(S)$, then $\theta =_\beta \sigma_S \circ \theta$, since $\theta(x_i) \xrightarrow{\ *\ }_\beta \theta(t_i) = \theta(\sigma_S(x_i))$ for $1 \leq i \leq n$, and $\theta(x) = \theta(\sigma_S(x))$ otherwise. Thus $\sigma_S \leq_\beta \theta$ and so obviously $\sigma_S \leq_\beta \theta[FV(S)]$. $\quad\square$

7.2.1 Transformations for Higher Order Unification

We may analyze the process of higher-order unification as follows. Let us assume, without loss of generality, that u and v are two lambda terms in $\mathcal{L}_{exp}$ and that θ is an idempotent, normalized unifier of u and v. Thus there exists some sequence of reductions to a β-normal form:

$$
\theta(u) \xrightarrow{\ *\ }_\beta w \xleftarrow{\ *\ }_\beta \theta(v).
$$

(Note that if all the terms instantiated by the substitution are first-order, then this sequence is trivial, since there are no β-reductions.) We may analyse this sequence *top-down*, examining the way in which each binding in the substitution (with its subsequent β-reduction, if the binding is higher-order) makes the two terms identical at each level of the terms. We have the following five cases (which are not intended to be mutually exclusive).

(A) $u = v$ and no unification is necessary. (Assume $u \neq v$ in the remaining cases.)

(B) No substitution takes place at the head in either term. In this case, $Head(u) = Head(v)$ and, since $u \neq v$, we must have $|u|, |v| > 0$. Thus, suppose $u = \lambda\overline{x_k}.\, a(\overline{u_n})$, $w = \lambda\overline{x_k}.\, a(\overline{w_n})$, and $v = \lambda\overline{x_k}.\, a(\overline{v_n})$, where $n > 0$ and either $a \in \Sigma$, or $a = x_i$ for some i, $1 \leq i \leq k$, or a is a free variable not in $D(\theta)$. In this case we must have

$$\theta(\lambda\overline{x_k}.\, u_i) \xrightarrow{\ *\ }_\beta \lambda\overline{x_k}.\, w_i \xleftarrow{\ *\ }_\beta \theta(\lambda\overline{x_k}.\, v_i)$$

for $1 \leq i \leq n$, that is, the subterms of u and v are pair-wise unifiable by θ.

(C) Our two terms are $u = \lambda\overline{x_k}.\, F(\overline{x_k})$ and $v = \lambda\overline{x_k}.\, v'$, for some variable F and some term v', and where $F \notin FV(v)$. In this case, we must have

$$\theta(\lambda\overline{x_k}.\, F(\overline{x_k})) \xleftarrow{\ *\ }_\beta \theta(\lambda\overline{x_k}.\, v'),$$

where $F \notin FV(v)$, and, if $\theta = [\lambda\overline{y_k}.t/F] \cup \theta'$, then since $\theta(F) \xleftarrow{\ }_\beta \theta(\lambda\overline{x_k}.\, F(\overline{x_k}))$,[6] we have $\theta(F) \xleftarrow{\ *\ }_\beta \theta(\lambda\overline{x_k}.\, v')$, where F does not occur in v', so that we may use the same argument we used in the first-order case. If we let $\sigma = [\lambda\overline{x_k}.\, v'/F]$ then $\theta =_\beta \sigma \circ \theta$, since θ and $\sigma \circ \theta$ differ only at F, but

$$\theta(F) \xleftarrow{\ *\ }_\beta \theta(\lambda\overline{x_k}.\, v') = \sigma \circ \theta(F).$$

This in fact shows that a pair of terms in this form has a single *mgu*. (For example, $\lambda x.\, F(x)$ and $\lambda x.\, f(x, z)$ are unified by $\theta = [\lambda y.\, f(y, a)/F, a/z]$, but $\sigma = [\lambda y.\, f(y, z)/F]$ is an *mgu*.) It should be obvious that this is a generalization of variable elimination to higher-order, since u is (up to η-equivalence) simply a variable not occurring in $FV(v)$.

(D) Some substitution takes place at the head of only one term; assume that this term is u (so that $Head(w) = Head(v)$). Then let $u = \lambda\overline{x_k}.\, F(\overline{u_n})$ and $v = \lambda\overline{x_k}.\, a(\overline{v_m})$ for some atom $a \neq F$ which is either a function constant, a bound variable, or a free variable not in $D(\theta)$. Now in order for the two terms to unify, we must make the head of u become a at some point in the sequence of β-reductions from $\theta(u)$ to w. There are two possibilities: either we imitate the head of v by substituting a term for F whose head is a, or we substitute a term for F which projects up a

---

[6] Note that the β-reduction simply replaces the bound variables $y_1, \ldots, y_k$ with $x_1, \ldots, x_k$, a useless operation in view of our assumption of α-equivalence.

subterm of u. (The latter case is only possible if F is of higher-order type.) We consider each of these in turn.

(Imitation) The substitution for F matches the head symbol of v by imitating the head symbol a, where $a \in \Sigma$ or a is a free variable not in $D(\theta)$, as we saw in Example 7.2.2.[7] Thus we have $\theta(F) = \lambda\overline{z_n}.\, a(\overline{r_m})$ for some terms $\overline{r_m}$ and we have a reduction sequence of the form

$$\theta(u) = \theta(\lambda\overline{x_k}.\, ((\lambda\overline{z_n}.\, a(\overline{r_m}))\overline{u_n})) \longrightarrow_\beta \theta(\lambda\overline{x_k}.\, a(\overline{r_m'})) \overset{*}{\longleftrightarrow}_\beta \theta(\lambda\overline{x_k}.\, a(\overline{v_m})),$$

where $r_i' = r_i[u_1/z_1, \ldots, u_n/z_n]$ for $1 \leq i \leq m$. (Notice that by the idempotency of θ, for illustration we can partially instantiate the term u with just the binding for the head F in this sequence.)

(Projection) The substitution for F attempts to match the head symbol a of v by projecting up a subterm of u. There are three ways to do this, depending upon the head symbol of the term projected up. First of all, perhaps a subterm of u has a head a which provides the match; for example, $F(\lambda x.\, f(x,a))$ and $f(b,a)$ will be unified by the substitution $[\lambda y.\, y(b)/F]$ in this fashion (note that we had to provide an argument b to the subterm $\lambda x.\, f(x,a)$ for the projection to work). The second reason to project is that perhaps a subterm of u is flexible, allowing us to start all over again in attempting to match the head of this new term to v. For example $F(\lambda x.\, G(x,a))$ and b can be unified by the substitution $[\lambda y.\, y(b)/F, \lambda x_1 x_2.\, x_1/G]$, where the binding for F works in this way. The third motivation for projection is that perhaps the subterm is itself a projection, and after some sequence of reductions, we have a term which is either flexible (and so we continue), or whose head is a and the match succeeds. For example, $\theta = [\lambda y_1.\, y_1(\lambda y_2.\, y_2(a))/F]$ unifies the two terms $u = F(\lambda x_1.\, x_1(\lambda x_2.\, f(x_2)))$ and $v = f(a)$ in this manner:

$$\begin{aligned}
\theta(u) &= [\lambda y_1.\, y_1(\lambda y_2.\, y_2(a))]\, \lambda x_1.\, x_1(\lambda x_2.\, f(x_2)) \\
&\longrightarrow_\beta\ [\lambda x_1.\, x_1(\lambda x_2.\, f(x_2))]\, \lambda y_2.\, y_2(a) \\
&\longrightarrow_\beta\ (\lambda y_2.\, y_2(a))\, \lambda x_2.\, f(x_2) \\
&\longrightarrow_\beta\ (\lambda x_2.\, f(x_2))\, a \\
&\longrightarrow_\beta\ f(a) = \theta(f(a)).
\end{aligned}$$

When we substitute a projection for the head of a flexible term $u = \lambda\overline{x_k}.\, F(\overline{u_n})$, we are restricted by the type of F to projecting up a subterm u_i which will preserve the type of u. In particular, since we can only

---

[7] Note that it is impossible to imitate a bound variable, since the rules of the calculus disallow free variable capture.

substitute a term of the same type as F, and since unification is only defined between terms of the same type, if $\tau(u) = \tau(v) = \alpha_1, \ldots, \alpha_k \to \beta$, then $\tau(u_i)$ must be some type $\gamma_1, \ldots, \gamma_{m'} \to \beta$ in order that the result of the projection preserves the type of u. Thus the type of the matrix of u_i must be the same as the matrix of u, and the substitution must provide arguments for each of the variables in the lambda binder of u_i. Thus if $\theta(F) = \lambda \overline{z_n}.\, z_i(\overline{r_{m'}})$ for some i, $1 \le i \le n$, then u_i must be in the form $u_i = \lambda \overline{y_{m'}}.\, u_i'$ where the type of the matrix u_i' is the same as the type of the matrices of u and v. In this case, the head a of u can be a function constant, a free variable, or a bound variable (i.e., one of the x_i), and thus we have a reduction sequence of the form

$$\theta(u) = \theta(\lambda \overline{x_k}.\, [(\lambda \overline{z_n}.\, z_i(\overline{r_{m'}}))\overline{u_n}]) \longrightarrow_\beta \theta(\lambda \overline{x_k}.\, [(\lambda \overline{y_{m'}}.\, u_i')\overline{r_{m'}'}])$$

$$\xrightarrow{\ *\ }_\beta \theta(\lambda \overline{x_k}.\, a'(\overline{t_p})) \xleftarrow{\ *\ }_\beta \theta(\lambda \overline{x_k}.\, a(\overline{v_m})),$$

where $r_i' = r_i[u_1/z_1, \ldots, u_n/z_n]$ for $1 \le i \le m'$,

$$\lambda \overline{x_k}.\, a'(\overline{t_p}) = (\lambda \overline{x_k}.\, [(\lambda \overline{y_{m'}}.\, u_i')\overline{r_{m'}'}]){\downarrow},$$

and either $a' = a$ or a' is a free variable in $D(\theta)$.

(E) Substitutions take place at the heads of both terms. Then let $u = \lambda \overline{x_k}.\, F(\overline{u_n})$ and $v = \lambda \overline{x_k}.\, G(\overline{v_m})$, where both F and G are in $D(\theta)$. Here we must eventually match the heads of the two terms, but we can do it in a large number of ways. In order to simplify our analysis, we attempt to reduce it to the previous case if we can. Let us (without loss of generality) focus on the binding made for the variable F. There are two subcases.

(i) θ substitutes a non-projection term for F, e.g., $\theta(F) = \lambda \overline{z_n}.\, a(\overline{s_p})$, where $a \ne G$ is not a bound variable (and by idempotency is not a variable in $D(\theta)$), and then (possibly) causes a β-reduction, after which we can analyse the result using case (D).

(ii) θ substitutes a projection term for F (which obeys the typing constraints discussed above), e.g., $\theta(F) = \lambda \overline{z_n}.\, z_j(\overline{t_q})$, and then, after we reduce to normal form, if the head symbol is either a function constant, a bound variable, or a variable not in $D(\theta)$, we may analyse the result using case (D); if the head is a variable in $D(\theta)$, then we (recursively) apply case (E) to these new terms.

By recursively applying this analysis to the subproblems generated we may account for every binding made by θ and every β-reduction in the original sequence. This forms the basis for the set of transformation rules below, which find unifiers by 'incrementally' building up bindings using partial bindings, as informally shown in the introduction. In case (D) above, this means that there will only be a finite number of choices for a partial binding, since there is only one possible imitation and only a finite number of possible projections. In case (E), unfortunately, this is not true. As shown in [72], the problem is that two flexible terms may not possess a finite CSU, and in fact there may be an infinite number of independent unifiers which contain flexible terms as bindings, so that even if we only attempt to find the top function symbol of the binding, there are potentially an infinite number of choices, since for each type there is always an infinite number of function variables. Thus, even if there is only a finite number of function constants in the language, it is not possible to reduce the non-determinism of this case in general to a finite number of choices of partial bindings, and so the search tree must be infinitely branching.[8]

Given a system S of terms from $\mathcal{L}_{exp}$ and some normalized $\theta \in U(S)$, a complete unification procedure must always be able to find some substitution σ such that $\sigma \in U(S)$ and $\sigma \leq_\beta \theta[FV(S)]$. Recall from the introduction that the basic idea of the transformation method is that, given some $\theta \in U(S)$, we attempt to find 'pieces' of θ by finding solved pairs x, t such that $\theta(x) \overset{*}{\longleftrightarrow}_\beta \theta(t)$; in this case, we know by an argument similar to that used in Lemma 7.2.7 that $\theta =_\beta [t/x] \circ \theta$, and by finding enough such pairs, we eventually have a $\sigma =_\beta [t_1/x_1] \circ \ldots \circ [t_n/x_n]$, where σ is a unifier of S more general than (or equivalent to) θ. In other words, we may successively approximate θ until we have built up just enough of the substitution to unify the system. We do this by 'solving' variables (as in case (C) above) or using approximations to individual bindings, as in Huet's method and in [54], which we call *partial bindings*.

Definition 7.2.8 A *partial binding of type* $\alpha_1, \ldots, \alpha_n \to \beta$ is a term of the form

$$\lambda \overline{y_n} . a\left(\lambda \overline{z_{p_1}^1} . H_1(\overline{y_n}, \overline{z_{p_1}^1}), \ldots, \lambda \overline{z_{p_m}^m} . H_m(\overline{y_n}, \overline{z_{p_m}^m})\right)$$

for some atom a, where

(1) $\tau(y_i) = \alpha_i$ for $1 \leq i \leq n$,
(2) $\tau(a) = \gamma_1, \ldots, \gamma_m \to \beta$, where $\gamma_i = \varphi_1^i, \ldots, \varphi_{p_i}^i \to \gamma_i'$ for $1 \leq i \leq m$,
(3) $\tau(z_j^i) = \varphi_j^i$ for $1 \leq i \leq m$ and $1 \leq j \leq p_i$;
(4) $\tau(H_i) = \alpha_1, \ldots, \alpha_n, \varphi_1^i, \ldots, \varphi_{p_i}^i \to \gamma_i'$ for $1 \leq i \leq m$.

The immediate subterms of a partial binding (i.e., the arguments to the atom a) will be called *general flexible terms*.

Note that these partial bindings are uniquely determined (up to renaming of the free variables) by their type and by their head symbol a.

Definition 7.2.9 For a partial binding as in the previous definition, if a is either a function constant or a free variable, then such a binding is called an *imitation binding for* a; if a is a bound variable y_i for some i, $1 \leq i \leq n$, then it is called an ith *projection binding*. A *variant* of a partial binding t is a term $\rho(t){\downarrow}$, where ρ is a renaming of the set $H_1, \ldots, H_m$ of free variables at the heads of the general flexible terms in t away from all variables in the context in which t will be used. For any variable F, a partial binding t is *appropriate to* F if $\tau(t) = \tau(F)$. An imitation binding is *appropriate to* $\lambda\overline{x_k}. F(\overline{u_n})$ iff it is appropriate to F.

In the case of an ith projection binding t for some i, $1 \leq i \leq n$, appropriate to a term $\lambda\overline{x_k}. F(\overline{u_n})$ of type $\alpha_1, \ldots, \alpha_k \to \beta$, the reader may check that $\tau(u_i) = \varphi_1, \ldots, \varphi_q \to \beta$ for some types $\varphi_1, \ldots, \varphi_q$, so that the result of substituting the binding and β-reducing will preserve the type of the term.

For notational brevity we shall extend our vector style notation to represent partial bindings in the form

$$\lambda\overline{y_n}. a(\overline{\lambda\overline{z_{p_m}}. H_m(\overline{y_n}, \overline{z_{p_m}})}).$$

Following our analysis of higher-order unification given above, we have the following set of transformations.

Definition 7.2.10 (The set of transformations $\mathcal{HT}$.) Let S be a system of lambda-terms (possibly empty). We have the following transformations.

(1) Trivial:

$$\{u, u\} \cup S \Longrightarrow S$$

(2) Term Decomposition: For any arbitrary atom a,

$$\{\lambda\overline{x_k}. a(\overline{u_n}), \lambda\overline{x_k}. a(\overline{v_n})\} \cup S \Longrightarrow \bigcup_{1 \leq i \leq n} \{\lambda\overline{x_k}. u_i, \lambda\overline{x_k}. v_i\} \cup S.$$

(3) Variable Elimination: If $u = \lambda\overline{x_k}. F(\overline{x_k})$ and $v = \lambda\overline{x_k}. v'$, for some k, some variable F, and some term v', where $F \notin FV(v)$, then

$$\{u, v\} \cup S \Longrightarrow \{F, \lambda\overline{x_k}. v'\} \cup \sigma(S){\downarrow},$$

where $\sigma = [\lambda\overline{x_k}. v'/F]$.

These three transformations are analogous to the set $\mathcal{ST}$. To provide for function variables, we need one more transformation, which is divided into three cases.

(4a) Imitation:

$$\{\lambda\overline{x_k}.\,F(\overline{u_n}),\,\lambda\overline{x_k}.\,a(\overline{v_m})\}\cup S \;\Longrightarrow\; \{F,\,t,\,\lambda\overline{x_k}.\,F(\overline{u_n}),\,\lambda\overline{x_k}.\,a(\overline{v_m})\}\cup S,$$

where a is either a function constant or a free variable not equal to F and where t is a variant of an imitation binding for a appropriate to F, e.g., $t = \lambda\overline{y_n}.\,a(\overline{\lambda\overline{z_{p_m}}.\,H_m(\overline{y_n},\overline{z_{p_m}})})$.

(4b) Projection:

$$\{\lambda\overline{x_k}.\,F(\overline{u_n}),\,\lambda\overline{x_k}.\,a(\overline{v_m})\}\cup S \;\Longrightarrow\; \{F,\,t,\,\lambda\overline{x_k}.\,F(\overline{u_n}),\,\lambda\overline{x_k}.\,a(\overline{v_m})\}\cup S,$$

where a is some arbitrary atom (possibly bound) and t is a variant of an ith projection binding for some i, $1 \le i \le n$, appropriate to the term $\lambda\overline{x_k}.\,F(\overline{u_n})$, that is, $t = \lambda\overline{y_n}.\,y_i(\overline{\lambda\overline{z_{p_q}}.\,H_q(\overline{y_n},\overline{z_{p_q}})})$, such that if $Head(u_i)$ is a function constant, then $Head(u_i) = a$.

(4c) Flex-Flex:

$$\{\lambda\overline{x_k}.\,F'(\overline{u_n}),\,\lambda\overline{x_k}.\,G(\overline{v_m})\}\cup S \;\Longrightarrow\; \{F,\,t,\,\lambda\overline{x_k}.\,F(\overline{u_n}),\,\lambda\overline{x_k}.\,G(\overline{v_m})\}\cup S,$$

where $t = \lambda\overline{y_n}.\,a(\overline{\lambda\overline{z_{p_m}}.\,H_m(\overline{y_n},\overline{z_{p_m}})})$ is a variant of some arbitrary partial binding appropriate to the term $\lambda\overline{x_k}.\,F(\overline{u_n})$ such that $a \ne F$ and $a \ne G$.

As a part of the transformations (4a)–(4c), we immediately apply Variable Elimination to the new pair F, t, which effectively amounts to just applying the substitution $[t/F]$ to the rest of the system. As in the other sets of transformations presented in this monograph, note that the unions above are *multiset unions*.

We shall say that Unify$(S) = \theta$ iff there exists a series of transformations $S \stackrel{*}{\Longrightarrow} S_n$, with S_n in solved form, and $\theta = \sigma_{S_n}|_{FV(S)}$.

Example 7.2.11 For example, the following series of transformations leads to a system in solved form.[9]

$$F(f(a)),\, f(F(a))$$
$$\Longrightarrow_{\text{imit}} F,\,\lambda x.\,f(Y(x)),\,\frac{(\lambda x.\,f(Y(x)))f(a)}{f(Y(f(a)))},\,f(\frac{(\lambda x.\,f(Y(x)))a}{f(Y(a))})$$
$$\Longrightarrow_{\text{dec}} F,\,\lambda x.\,f(Y(x)),\,Y(f(a))),\,f(Y(a))$$
$$\Longrightarrow_{\text{proj}} F,\,\lambda x.\,f(\frac{(\lambda x.\,x)x}{x}),\,Y,\,\lambda x.\,x,\,\frac{(\lambda x.\,x)f(a)}{f(a)},\,f(\frac{(\lambda x.\,x)a}{a})$$
$$\Longrightarrow_{\text{triv}} F,\,\lambda x.\,f(x),\,Y,\,\lambda x.\,x$$

Hence, $[\lambda x.\,f(x)/F]$ is a unifier of the original two terms.

---

[9] In order to show the effect of the β-reductions which follow the application of substitutions in (3), we often explicitly represent these reductions using an 'inference' style notation, e.g., we represent the effect of the substitution θ on the term e as $\frac{\theta(e)}{\theta(e)\downarrow}$, to illustrate both the effect of the substitution and the subsequent β-normal form.

7.2.2 Soundness of the Transformations

The following lemmas will enable us to prove the soundness of this set of transformations.

Lemma 7.2.12 If $S \implies S'$ using Trivial or Variable Elimination, then $U(S) = U(S')$.

Proof. As in the first-order case, the only difficulty is in Variable Elimination. We must show that $U(\{x, v\} \cup S) = U(\{x, v\} \cup \sigma(S)\!\downarrow)$ where $\sigma = [v/x]$ and $x \notin FV(v)$. For any substitution θ, if $\theta(x) \overset{*}{\longleftrightarrow}_\beta \theta(v)$, then $\theta =_\beta \sigma \circ \theta$, since $\sigma \circ \theta$ differs from θ only at x, but $\theta(x) \overset{*}{\longleftrightarrow}_\beta \theta(v) = \sigma \circ \theta(x)$. But then, using Lemma 7.1.21, it is easy to see that $\theta \in U(S)$ iff $\sigma \circ \theta \in U(S)$. Furthermore, since for any term u we must have $\sigma \circ \theta(u) = \theta(\sigma(u)) \overset{*}{\longrightarrow}_\beta \theta(\sigma(u)\!\downarrow)$, it can easily be shown that $\sigma \circ \theta \in U(S)$ iff $\theta \in U(\sigma(S)\!\downarrow)$. Thus,

$$\theta \in U(\{x, v\} \cup S)$$
$$\text{iff} \quad \theta(x) \overset{*}{\longleftrightarrow}_\beta \theta(v) \text{ and } \theta \in U(S)$$
$$\text{iff} \quad \theta(x) \overset{*}{\longleftrightarrow}_\beta \theta(v) \text{ and } \sigma \circ \theta \in U(S)$$
$$\text{iff} \quad \theta(x) \overset{*}{\longleftrightarrow}_\beta \theta(v) \text{ and } \theta \in U(\sigma(S)\!\downarrow)$$
$$\text{iff} \quad \theta \in U(\{x, v\} \cup \sigma(S)\!\downarrow).$$

$\square$

This lemma shows that the invariant properties of a problem are preserved under these two transformations, as they were in the first-order case.

Lemma 7.2.13 Suppose that $S \implies_{\text{dec}} S'$ where the pair in S transformed is $\lambda \overline{x_n}. a(\overline{u_n}), \lambda \overline{x_n}. a(\overline{v_n})$. For any substitution θ,
(i) if a is either a constant or a bound variable or a free variable *not* in $D(\theta)$, then $\theta \in U(S)$ iff $\theta \in U(S')$;
(ii) if $a \in D(\theta)$ then $\theta \in U(S')$ implies that $\theta \in U(S)$.

Proof. If $\theta(\lambda \overline{x_k}. u_i) \overset{*}{\longleftrightarrow}_\beta \theta(\lambda \overline{x_k}. v_i)$ for $1 \leq i \leq n$, then clearly we must have

$$\theta(\lambda \overline{x_k}. a(\overline{u_n})) = \lambda \overline{x_k}. \theta(a)(\theta(u_1), \ldots, \theta(u_n)) \overset{*}{\longleftrightarrow}_\beta$$
$$\lambda \overline{x_k}. \theta(a)(\theta(v_1), \ldots, \theta(v_n)) = \theta(\lambda \overline{x_k}. a(\overline{v_n})),$$

and so for any atom a we have $\theta \in U(S)$ whenever $\theta \in U(S')$. If a is either a function constant, a bound variable, or a variable not in $D(\theta)$, then $\theta(a) = a$ and it is easy to see that the reverse direction holds as well.

$\square$

Lemma 7.2.14 If $S \implies S'$ using Term Decomposition, Imitation, Projection, or Flex-Flex then $U(S') \subseteq U(S)$.

Proof. For Term Decomposition the result is a consequence of our previous lemma. The remaining transformations are two parts, first adding a pair F, t to the system S, and then applying Variable Elimination to this new pair. Clearly, since $S \subseteq \{F, t\} \cup S$ we must have $U(\{F, t\} \cup S) \subseteq U(S)$. That the subsequent application of Variable Elimination to the new pair is sound has been shown by Lemma 7.2.12. $\square$

Since in the last three transformations we effectively commit ourselves to a particular approximation of a solution, it is hardly surprising that the inclusion $U(S') \subseteq U(S)$ is in general proper. Similarly, in the case of Term Decomposition, decomposing flexible pairs may eliminate unifiers; for example $F(a, b), F(c, d)$ has an infinite number of unifiers, but the system a, c, b, d has none. These results show us that in higher-order unification, the set of solutions is invariant only under Trivial, Variable Elimination, and under Term Decomposition in the case of two rigid terms.

Finally, using these lemmas we have

Theorem 7.2.15 (Soundness) If $S \overset{*}{\implies} S'$, with S' in solved form, then the substitution $\sigma_{S'}|_{FV(S)} \in U(S)$.

Proof. By a simple induction on the length of transformation sequences, and using the previous lemmas in the induction step, we may show that $\sigma_{S'} \in U(S)$. But since the restriction has no effect as regards the effect of the substitution on the terms in S, we see that $\sigma_{S'}|_{FV(S)} \in U(S)$. $\square$

7.2.3 Completeness of the Transformations

The completeness of our set of transformations will be proved in a manner analogous to that used for the proof of completeness of the set $\mathcal{ST}$, except that now the transformation relation is not terminating in general, so we shall prove only the *non-deterministic completeness* of the set, i.e., we show that for any system S, if $\theta \in U(S)$, then there exists *some* sequence of transformations which finds a unifier σ such that $\sigma \leq_\beta \theta[FV(S)]$.

First we show the exact sense in which partial bindings can be considered to be approximations to bindings in substitutions.

Lemma 7.2.16 If $s = \lambda \overline{x_n}. a(\overline{s_m})$ is any term, then there exists a variant of a partial binding t and a substitution η such that $\eta(t) \overset{*}{\longrightarrow}_\beta s$.

Proof. If $m = 0$, i.e. $s = \lambda \overline{x_k} . a$, then the result is trivial by taking $t = s$ and $\eta = Id$. Otherwise, assume $m > 0$, and let

$$t = \lambda \overline{x_n} . a(\overline{\lambda \overline{z_{p_m}} . H_m(\overline{x_n}, \overline{z_{p_m}})})$$

and $\eta = [\lambda \overline{x_n} . s_1 / H_1, \ldots, \lambda \overline{x_n} . s_m / H_m]$. Then by the type of the head a, the ith subterm s_i must be in the form $\lambda z_{p_i} . s_i'$, so that

$$\eta(\lambda \overline{z_{p_i}} . H_i(\overline{x_n}, \overline{z_{p_i}})) \longrightarrow_\beta s_i,$$

for each i, $1 \le i \le m$. Thus $\eta(t) \stackrel{*}{\longrightarrow}_\beta s$. $\square$

Lemma 7.2.17 If $\theta = [s/F] \cup \theta'$ then there exists a variant of a partial binding t appropriate to F and a substitution η such that

$$\theta = [s/F] \cup \eta \cup \theta'[D(\theta)]$$
$$=_\beta [t/F] \circ \eta \cup \theta'[D(\theta)].$$

Furthermore, if $D(\theta) \cap I(\theta) = \emptyset$, then $\theta'' = [s/F] \cup \eta \cup \theta'$ is a unifier of the pair F, t and $D(\theta'') \cap I(\theta'') = \emptyset$.

Proof. Given the term s, let t and η be as in the previous lemma. Since t is a variant, $D(\eta) \cap D(\theta) = \emptyset$, and since furthermore $\eta(t) \stackrel{*}{\longrightarrow}_\beta s$, we have $[s/F] = [s/F] \cup \eta =_\beta [t/F] \circ \eta[D(\theta)]$, from which the first part follows. If $D(\theta) \cap I(\theta) = \emptyset$ (so that θ is idempotent), then since t is a variant, $D(\eta) \cap I(\theta) = \emptyset$, so that $D(\theta'') \cap I(\theta'') = \emptyset$ (and $\theta''(s) = s$) and finally, $\theta''(F) = s \stackrel{*}{\longleftarrow}_\beta \eta(t) = \theta''(t)$. $\square$

Note that if $D(\theta) \cap I(\theta) \neq \emptyset$ in this lemma, then potentially θ has a binding for the head of s and t, and so possibly $\theta''(t) \neq \eta(t)$. Also, notice that $[s/F] \cup \eta$ and $[t/F] \circ \eta$ are only β-equal (over $D(\theta)$) because we do not assume that the implicit β-reductions are performed when substitutions are composed. These lemmas show the motivation for the term 'partial binding' and provide the formal justification for the assertion that partial bindings can be used to build up substitutions incrementally.

Next we define a set of transformations on pairs θ, S which shows how the structure of a substitution θ can determine an appropriate sequence of transformations. For simplicity here we shall refer to the transformations in $\mathcal{HT}$ by number instead of name.

Definition 7.2.18 (The set $\mathcal{CT}$) Let θ be a normalized substitution and S be an arbitrary system. The first three transformations are essentially from the set $\mathcal{HT}$:

$$\theta, S \implies_i \theta, S'$$

for $1 \leq i \leq 3$ iff $S \implies_i S'$ in the set $\mathcal{HT}$, with the restriction that (2) is only applied to a pair u, v if the top function symbol in u and v is *not* a free variable in $D(\theta)$. Also, we have

$$[s/F] \cup \theta, \{\lambda \overline{x_k}.\, F(\overline{u_n}), \lambda \overline{x_k}.\, v\} \cup S \implies_4$$
$$[s/F] \cup \eta \cup \theta, \{F, t, \lambda \overline{x_k}.\, F(\overline{u_n}), \lambda \overline{x_k}.\, v\} \cup S,$$

where F is not solved in the system on the left side, s is some term $\lambda \overline{y_n}.\, a(\overline{s_m})$,

$$t = \lambda \overline{y_n}.\, a(\overline{\lambda \overline{z_{p_m}}.\, H_m(\overline{y_n}, \overline{z_{p_m}})})$$

is a partial binding appropriate to F with the same (up to α-conversion) head as s, and

$$\eta = [\lambda \overline{y_n}.\, s_1/H_1, \ldots, \lambda \overline{y_n}.\, s_m/H_m].$$

(Note that perhaps $m = 0$ in which case η is omitted.) Transformation (3) is immediately applied as a part of (4), as in the set $\mathcal{HT}$. Again, notice that $[s/F] \cup \eta \cup \theta =_\beta [t/F] \circ \eta \cup \theta$.

Example 7.2.19 Let $\theta = [\lambda x.\, f(x)/F]$ and $S = \{F(f(a)), f(F(a))\}$. We have the following sequence of $\mathcal{CT}$-transformations.

$[\lambda x.\, f(x)/F], \{F(f(a)), f(F(a))\}$
$\implies_4 [\lambda x.\, f(x)/F, \lambda x.\, x/Y], \{F, \lambda x.\, f(Y(x)), \frac{(\lambda x.\, f(Y(x)))f(a)}{f(Y(f(a)))}, f(\frac{(\lambda x.\, f(Y(x)))a}{f(Y(a))})\}$
$\implies_2 [\lambda x.\, f(x)/F, \lambda x.\, x/Y], \{F, \lambda x.\, f(Y(x)), Y(f(a)), f(Y(a))\}$
$\implies_4 [\lambda x.\, f(x)/F, \lambda x.\, x/Y], \{F, \lambda x.\, f(\frac{(\lambda x.\, x)x}{x}), Y, \lambda x.\, x, \frac{(\lambda x.\, x)f(a)}{f(a)}, f(\frac{(\lambda x.\, x)a}{a})\}$
$\implies_1 [\lambda x.\, f(x)/F, \lambda x.\, x/Y], \{F, \lambda x.\, f(x), Y, \lambda x.\, x\}$

The next lemma shows us how these transformations are useful for proving completeness.

Lemma 7.2.20 If $\theta \in U(S)$ for some system S not in solved form, and W is a set of variables, then there exists some transformation $\theta, S \implies \theta', S'$ such that
 (i) $\theta = \theta'[W]$;
 (ii) If $D(\theta) \cap I(\theta) = \emptyset$ then $\theta' \in U(S')$ and $D(\theta') \cap I(\theta') = \emptyset$; and
 (iii) $S \implies S'$ with respect to the set $\mathcal{HT}$.

Proof. Since S is not in solved form, there must exist some pair u, v which is not solved in S. We have three cases: (A) If $u = v$ then we may apply (1) or (2); (B) if $Head(u) = Head(v) \notin D(\theta)$, then we can apply (2); otherwise, (C) we have $u \neq v$ and either $Head(u) \neq Head(v)$ or $Head(u) = Head(v) \in D(\theta)$. In case (C), either u or v has an unsolved variable from $D(\theta)$ at its head; without loss of generality, assume that u has. Thus, we have $u = \lambda \overline{x_k}. F(\overline{u_n})$ and $v = \lambda \overline{x_k}. v'$ with $F \in D(\theta)$ and F not solved in S and (4) must apply, and in the special case that $u \xrightarrow{\ *\ }_\eta F$ and $F \notin FV(v)$, we can alternately apply (3). Although there may not be a unique choice about which transformation to apply, at least one must apply, and thus we have some transformation $\theta, S \Longrightarrow_i \theta', S'$. In the case that $1 \leq i \leq 3$, (i) holds because $\theta' = \theta$, by our soundness lemmas of the previous section we have (ii), and (iii) holds by the definition of the set CT. If $i = 4$ then by our previous corollary we have extended $\theta = [s/F] \cup \varphi$ to a substitution $\theta' = [s/F] \cup \eta \cup \varphi =_\beta [t/F] \circ \eta \cup \varphi$ where we can assume that $D(\eta) \cap W = \emptyset$ (showing (i)), and we have added a pair F, t to S to form S'. From the definition of CT and the previous lemma it is clear that we have $D(\theta') \cap I(\theta') = \emptyset$ and $\theta'(F) = s \xleftarrow{\ *\ }_\beta \eta(t) = \theta'(t)$, so that $\theta' \in U(S')$, showing (ii). Finally, since S is unifiable it is not hard to see that the conditions imposed on (4) in CT are consistent with (4) in $\mathcal{HT}$. If $Head(v)$ is not a variable in $D(\theta)$, then we have two cases: if $Head(s) = Head(v)$, then $S \Longrightarrow_{4a} S'$ (i.e., this is an imitation case); otherwise, s is a projection, and $S \Longrightarrow_{4b} S'$. If $Head(v) \in D(\theta)$ then $S \Longrightarrow_{4c} S'$. $\square$

Corollary 7.2.21 If $\theta \in U(S)$ and no transformation applies to θ, S then S is in solved form.

Finally, we may present our completeness proof.

Theorem 7.2.22 (Completeness of $\mathcal{HT}$) For any system S, if $\theta \in U(S)$ then there exists some sequence of transformations

$$S = S_0 \implies S_1 \implies S_2 \implies \ldots \implies S_n,$$

where S_n is in solved form and $\sigma_{S_n} \leq_\beta \theta[FV(S)]$.

Proof. By Lemma 7.1.26 we may assume without loss of generality that $D(\theta) \cap I(\theta) = \emptyset$ (since if not we may find a substitution $\theta'' = \theta[FV(S)]$ fulfilling these conditions). We prove this result using the set CT, first showing that every sequence of CT transformations terminates. For any θ and S, define the complexity measure $\mu(\theta, S) = \, <M, n>$, where n is the sum of the sizes (i.e., the number of subterms) of all terms in S, and M is a multiset of integers corresponding to the sizes of the bindings in θ for variables which are not solved in S:

$$M = \{\, |\theta(x)| \mid x \in D(\theta) - Sol(S) \,\},$$

where $Sol(S)$ is the set of all variables solved in S. The lexicographic ordering on pairs which uses the standard multiset ordering for the first component and $<$ on the natural numbers for the second is well-founded and any $\mathcal{CT}$-transformation produces a pair strictly smaller under the ordering: (1) and (2) reduce n without affecting M, (3) reduces M by removing a variable from $D(\theta) - Sol(S)$, and (4) reduces M by removing a variable from $D(\theta) - Sol(S)$ and replacing it with some number of new, unsolved variables whose bindings in θ' are all smaller than the binding removed (since they are proper subterms of it). Hence every sequence of $\mathcal{CT}$-transformations is finite.

Thus there must exist a sequence of transformations

$$\theta, S = \theta_0, S_0 \implies \theta_1, S_1 \implies \ldots \implies \theta_m, S_m$$

such that no transformation applies, and by induction on m using the previous lemma, with $FV(S)$ for the set W, we have $\theta = \theta_m[FV(S)]$, $\theta_m \in U(S_m)$, and there is a corresponding sequence of $\mathcal{HT}$-transformations

$$S = S_0 \implies S_1 \implies \ldots \implies S_m$$

and by the corollary we know that S_m is in solved form. Finally, by Lemma 7.2.7 we have $\sigma_{S_m} \leq_\beta \theta_m = \theta[FV(S)]$. $\quad\square$

The reader should note that this proof is essentially similar to that of Theorem 3.3.9. Finally, combining our soundness and completeness results, we have that this method is capable of non-deterministically finding a unifier of S more general than any given unifier. More formally, we may characterize the set of substitutions non-deterministically found by the set of transformations $\mathcal{HT}$ as follows.

Theorem 7.2.23 For any system S, the set

$$\{\sigma_{S'}|_{FV(S)} \mid S \overset{*}{\implies} S', \text{ and } S' \text{ is in solved form}\}$$

is a $CSU(S)$. By application of the appropriate renaming substitution away from W, this set is a $CSU(S)[W]$ for any W.

Proof. We must simply verify the conditions in Definition 7.2.5. Coherence was shown in Theorem 7.2.15 and our previous result demonstrated completeness. By restricting the idempotent substitution $\sigma_{S'}$ to $FV(S)$ we satisfy purity for W empty. If W is not empty, we may suitably rename the variables introduced by each of the substitutions $\sigma_{S'}$ away from W, using Lemma 7.2.6. □

The careful reader will note that we have made no assumptions about the order in which transformations are performed, and so these results apply in a very general way to the derivation of solved form systems from initial systems of equations. In particular, we see that the strategy of *eager variable elimination*, in which transformation Variable Elimination is performed as soon as possible on any pair to which it applies, is complete (in the case of general E-unification this problem is still open, as discussed in section §6.7). The search space is thereby reduced, since we do not need to build up such solved pairs one symbol at a time. In addition, it shows how this set of transformations is a true generalization of the transformations used for first-order unification.

7.3 Huet's Procedure Revisited

The set of transformations given in the previous section were proved to be complete for the problem of general higher-order unification, that is, they can non-deterministically find *any* higher-order unifier of two arbitrary terms. Unfortunately, as remarked above, the 'don't know' non-determinism of this set causes severe implementation problems in the case of two flexible terms (case (E) in our analysis), and, as discussed above, this 'guessing' of partial bindings in this case can not be avoided without sacrificing completeness, and so the search tree of all transformation sequences may be infinitely branching at certain nodes, causing a disastrous explosion in the size of the search space.

Huet's well-known solution to this problem [71,72] was to redefine the problem in such a way that such flexible-flexible pairs are considered to be already solved; this partial solution of the general higher-order unification problem turns out to be sufficient for refutation methods (see [69]), and this is the method used in most current systems. We show here how to explain this approach in terms of transformations on systems. The only changes have to do with redefining the notion of a solved system and restricting the set of transformations.

Definition 7.3.1 A pair of terms x, e is in *presolved form* in a system S if it is in solved form in S (as above) *or* if it is a pair consisting of two flexible terms. A system is in presolved form if each member is in presolved form. For a set S in presolved form, define the associated substitution σ_S as the *mgu* $\sigma_{S'}$ of the set S' of solved pairs of S.

Definition 7.3.2 Let $\cong$ be the least congruence relation on $\mathcal{L}$ containing the set of pairs $\{(u, v) \mid u, v \text{ are both flexible terms }\}$. A substitution θ is a *preunifier* of u and v if $\theta(u){\downarrow} \cong \theta(v){\downarrow}$.

The importance of pre-unifiers is shown by our next definition and lemma.

Definition 7.3.3 For every $\phi = \alpha_1, \ldots, \alpha_n \to \beta \in \mathcal{T}$, with $n \geq 0$, define a term

$$\widehat{e}_\phi = \lambda x_1 \ldots x_n . v,$$

where $\tau(x_i) = \alpha_i$ for $1 \leq i \leq n$ and $v \in V_\beta$ is a new variable which will never be used in any other term. Let ζ be an (infinite) set of bindings

$$\zeta = \{\widehat{e}_{\tau(x)}/x \mid x \in V\}.$$

Finally, if S' is a pre-solved system containing a set S'' of flexible-flexible pairs, then define the substitution

$$\zeta_{S'} = \zeta|_{FV(S'')}.$$

As in [72], it is easy to show this next result.

Lemma 7.3.4 If S is a system in pre-solved form then the substitution $\sigma_S \cup \zeta_S$ is a unifier of S.

This lemma asserts that pre-unifiers may always be extended to true unifiers by finding trivial unifiers for the flexible-flexible terms in the pre-solved system.

The set of transformations for finding preunifiers is a slightly restricted version of the set of transformations $\mathcal{HT}$.

Definition 7.3.5 (The set of transformations $\mathcal{PT}$) Let S be a system, possibly empty. To the transformations Trivial and Variable Elimination from $\mathcal{HT}$ we add three (restricted) transformations.

Transformation $(2')$ is

$$\left\{\lambda\overline{x_k}. a(\overline{u_n}), \lambda\overline{x_k}. a(\overline{v_n})\right\} \cup S \Longrightarrow \bigcup_{1 \leq i \leq n} \left\{\lambda\overline{x_k}. u_i, \lambda\overline{x_k}. v_i\right\} \cup S,$$

where $a \in \Sigma$ or $a = x_j$ for some j, $1 \leq j \leq k$. For $(4'a)$ we have

$$\{\lambda \overline{x_k}. F(\overline{u_n}), \lambda \overline{x_k}. a(\overline{v_m})\} \cup S \implies \{F, t, \lambda \overline{x_k}. F(\overline{u_n}), \lambda \overline{x_k}. a(\overline{v_m})\} \cup S,$$

where $a \in \Sigma$ and t is a variant of an imitation binding for a appropriate to F. Finally, we define $(4'b)$ as

$$\{\lambda \overline{x_k}. F(\overline{u_n}), \lambda \overline{x_k}. a(\overline{v_m})\} \cup S \implies \{F, t, \lambda \overline{x_k}. F(\overline{u_n}), \lambda \overline{x_k}. a(\overline{v_m})\} \cup S,$$

where either $a \in \Sigma$ or $a = x_j$ for some j, $1 \leq j \leq k$, and t is a variant of an ith projection binding for some i, $1 \leq i \leq n$, appropriate to the term $\lambda \overline{x_k}. F(\overline{u_n})$.

After each of $(4'a)$ and $(4'b)$, we apply Variable Elimination to the new pair introduced. As in our previous definitions, recall that the unions are multiset unions.

We shall say that Pre-Unify$(S) = \theta$ iff there exists a series of transformations from $\mathcal{PT}$

$$S = S_0 \implies S_1 \implies \ldots \implies S_n,$$

with S_n in pre-solved form, and $\theta = \sigma_{S_n}|_{FV(S)}$.

In terms of Huet's procedure (see the Appendix) the first two transformations represent approximately the effect of Simplify, and $(4'a)$ and $(4'b)$ represent the processes of imitation and projection respectively in Match. Variable Elimination represents the effect of applying substitutions in Simplify, but also more generally allows solving certain pairs immediately, which was remarked upon by Huet (see [72], p. 3-57) but not emphasized.[10] Note that the transformations Trivial, $(2')$, and Variable Elimination in $\mathcal{PT}$ preserve the set of solutions invariant, as discussed in Section §7.2.2.

We now present the major results concerning this formulation of higher-order unification, following [72]. Their proofs are simple modifications of our previous results, and are omitted.

Theorem 7.3.6 (Soundness) If $S \overset{*}{\implies} S'$, with S' in presolved form, then the substitution $\sigma_{S'}|_{FV(S)}$ is a preunifier of S.

Theorem 7.3.7 (Completeness) If θ is some preunifier of the system S, then there exists a sequence of transformations $S \overset{*}{\implies} S'$, with S' in presolved form, such that

$$\sigma_{S'}|_{FV(S)} \leq_\beta \theta.$$

---

[10] Jensen and Pietrzykowski [129] suggest a similar rule as a heuristic improvement.

The search tree for this method consists of all the possible sequences of systems created by transforming the original two terms. Leaves consist of pre-solved systems or systems where no transformation can be applied. These correspond to the **S** and **F** nodes in Huet's algorithm; in fact, the search trees generated are essentially the same as the *matching trees* defined in [71], except that here an explicit representation of the matching substitutions found so far is carried along in the system (see the Appendix). The set of pre-unifiers potentially found by our procedure is the set of pre-solved leaves in the search tree.

As in the case of general higher-order unification, the strategy of eager variable elimination is complete, allowing a reduction in the size of the search space, since we do not need to build up the terms using partial bindings. This rule had been suggested as a heuristic in [72] and [129], but not emphasized as an essential part of the method of building up substitutions, as here. We note also as a minor point that in some cases it is possible to apply variable elimination to a presolved system so that that this binding is incorporated into the *mgu* of the final solved form system. For example, the following initial system is presolved, but in fact has a *mgu* $[\lambda x. G(a, x)/F]$:

$$\lambda x. F(x), \lambda x. G(a, x), F(b), G(a, b)$$
$$\Longrightarrow_3 F, \lambda x. G(a, x), \frac{(\lambda x. G(a,x))\, b}{G(a,b)}, G(a, b)$$
$$\Longrightarrow_1 F, \lambda x. G(a, x).$$

We give a pseudo-code version of Huet's method for the typed $\beta\eta$-calculus in Appendix Two as an example of the way in which these transformations can be used to design more practical procedures.

7.4 Conclusion

We have presented in this chapter a reexamination of the problem of general higher-order unification, using the abstract approach of transformations on systems of equations. As in the case of E-unification, this kind of analysis provides the right level of abstraction by revealing the logical issues in their purest form. We claim that this approach is more perspicuous than those previously advanced, permits more direct soundness and completeness proofs, and unifies and justifies the various approaches taken to unification problems. This abstract characterization of the process of unification in various settings clarifies the basic similarities and differences of the problems by removing the notion of control and showing exactly where non-determinism arises and where it may be eliminated. The three sets of transformations $\mathcal{ST}$, $\mathcal{PT}$, and $\mathcal{HT}$ thus represent an (inclusion) hierarchy of abstract methods for unification. One new result that came out of this is that variable elimination can be extended from first-order unification to both general higher-order unification and to pre-unification; in particular, the strategy of eager variable elimination is still complete.

CHAPTER 8

CONCLUSION

In this monograph we studied general E-unification and higher-order unification using the method of non-deterministic transformations on systems of equations originated by Herbrand and developed in the case of standard first-order unification by Martelli and Montanari. This formalism provides an abstract and mathematically elegant means of analysing the properties of these more complex types of unification problems by providing a clean separation of the logical issues from the specification of procedural information. In each case, we extended the basic set of transformations ST for standard unification by analysing the precise manner in which terms are defined to be 'the same' in these two generalizations of unification, i.e., modulo the least congruence induced by the set of equations for E-unification, and modulo the conversion rules of the typed lambda calculus for higher-order unification.

In the case of E-unification, we extended the basic set of transformations ST for standard unification to two sets of transformations BT and T which are sound and complete for E-unification in arbitrary equational theories. A new set T' may be obtained by refining the set T even further. This section of the book provides the first presentation of an E-unification procedure complete for arbitrary sets of equations.

In the higher-order case, our major contribution was the presentation of an abstract and simplified non-deterministic method for general higher-order unification of which first-order unification and higher-order preunification are special cases, a more direct proof of completeness, and a proof that the strategy of *eager variable elimination* in this context is complete. This set of transformations is derived from an analysis of the role of substitution and β-reduction in unification, which we feel clarifies the design of Huet's procedure, and shows how its basic principles work in a more general setting. We claim that this approach is more perspicuous than those previously advanced, and unifies and justifies the various approaches taken to unification problems. The three sets of transformations ST, PT, and HT thus represent an (inclusion) hierarchy of abstract methods for unification.

Since this research project was first undertaken, the techniques developed here have been applied with success to a new and very powerful form of unification, namely, *higher-order E-unification*. In this problem, given two higher-order terms e_1 and e_2 and a set of first-order equations E, we ask whether there exists a substitution θ such that $\theta(e_1) \xleftrightarrow{\ *\ }_{\beta\eta E} \theta(e_2)$, where $\longleftrightarrow_{\beta\eta E}$ is the union of $\longleftrightarrow_{\beta\eta}$ and $\longleftrightarrow_E$. The results of Breazu-Tannen and Gallier [23] essentially show that the higher-order and the first-order equational parts of the problem interact smoothly and without pathologies, and so the results presented here in Chapters §6 and §7 can be combined. These results were presented by the author in preliminary for in [150]. In addition, the method of transformations has been used to show that the higher-order E-unification problem can be modularized to a large extent into a higher-order unification module and an E-unification module, under certain conditions, by Nipkow and Qian [119], and to show that higher-order unification and higher-order E-unification can be developed in the context of combinatory logic [40,81]. We might also mention several surveys on unification which have appeared recently and which present their subject in the formalism of transformations; two by J. Gallier and the author [55,56] which present the results of [53] and [148] (which form the basis for this monograph), in abbreviated form, plus a presentation of Rigid E-Unification [52]; the reader should also be aware of the recent survey by Jouannaud and Kirchner [83].

The results presented in this monograph, and the recent developments outlined above, show that the formalism of non-deterministic transformations on systems of equations—which is simply an inference system for unification—provides the right level of abstraction to form the basis of a proof theory of general unification by revealing the logical issues in their purest form. It is our hope that this formalism, in addition to providing a theoretical foundation both for the study of general unification methods in theorem proving and logic programming, will provide a unifying connection between the diverse approaches to E-unification and higher order unification currently being developed and the larger concerns of the proof theory of mathematical logic.

Appendix 1

DETERMINISTIC E-UNIFICATION

In this appendix, we design a deterministic procedure by emphasizing a distinction implicit in $\mathcal{BT}$, viz., whether a rewrite takes place at the root (Root Rewriting) or not (the other transformations). Specifically, if u and v are E-unifiable, then $\forall \theta \in U_E(u, v)$ there exists a sequence $\theta(u) = u_0 \longleftrightarrow_E u_1 \longleftrightarrow_E \ldots \longleftrightarrow_E u_n = \theta(v)$. Each such θ can be classified into at least one, and possibly both, of the following two cases. (Case 1 is further divided into five mutually exclusive cases based on the stucture of the terms.)

1. No rewrite rule is applied at the root of any u_i.
   (a) Both u and v are compound terms, e.g., $u = f(u_1, \ldots, u_n)$ and $v = f(v_1, \ldots, v_n)$. Thus $\theta(u_i) \overset{*}{\longleftrightarrow}_E \theta(v_i)$ for $1 \leq i \leq n$.
   (b) Either u or v is a variable; assume u is a variable.
      i. v is a constant or a variable.
      ii. v is a compound term.
         A. $u \notin Var(v)$.
         B. $u \in Var(v)$, so if $v = f(v_1, \ldots, v_n)$ then $\theta(u) = f(t_1 \ldots t_n)$ for some terms $t_1, \ldots, t_n$.
   (c) Both u and v are constants, i.e., $u = v$.
2. Some rewrite rule is applied at the root of some u_i. Thus

$$\theta(u) \overset{*}{\longleftrightarrow}_E \rho(l) \longleftrightarrow_E \rho(r) \overset{*}{\longleftrightarrow}_E \theta(v),$$

where $(l \doteq r)$ is a variant of an equation in $E \cup E^{-1}$, ρ is the matching substitution used in the rewrite step, and no root rewrite takes place between $\theta(u)$ and $\rho(l)$.

The following Pseudo-Pascal procedure recursively applies this classification to two terms, adapting the control strategy of Robinson's original algorithm for standard unification [139] to the case of E-unification, and using depth-first iterative deepening to simulate breath-first search.

```
global variables
  currDepth, maxDepth : integer;   E : eqSet;

procedure E-Unifiers( u, v : term );
  begin
    for maxDepth := 1 to ∞ do
      begin
        currDepth := 0;
        output( E-Unifs(u, v, false, false ) )
      end
  end;

function E-Unifs( u, v : term; occur, noRootRW : boolean ) : unifSet;
var
  unifs1, unifs2, subUnifs : unifSet;   i, n : integer;   θ, σ : unifier;
  f : funcSymbol; y₁,...,yₙ : variables;
begin
  currDepth := currDepth + 1;
  if currDepth > maxDepth
    then return(∅);                    { Terminate this call and return ∅ }

{ Case 1:   Find unifiers of u and v which don't rewrite root and collect in unifs1 }

  if u = v
    then unifs1 := {Id}            { This includes Case 1.(c) }
  else if |u| > 0 and |v| > 0 and (Root(u)=Root(v))      { Case 1.(a) }
    then begin
      unifs1 := E-Unifs( u/1, v/1, false, false );
      for i := 2 to Arity(Root(u)) do
        begin
          subUnifs := ∅;
          for each θ ∈ unifs1 do
            subUnifs := subUnifs ∪ θ ∘ E-Unifs(θ(u/i), θ(v/i), false, false);
          unifs1 := subUnifs
        end
    end
  else if Variable(u) or Variable(v)
    then begin
      if not Variable(u)
        then Swap(u, v);
      if |v| = 0 or (|v| > 0 and (u ∉ Vars(v)))
        then unifs1 := {[v/u]}          { Cases 1.(b).i to 1.(b).ii.B }
        else begin             { Case 1.(b).ii.B }
          if not occur           { start of new occur check case found }
            then Mark all addresses α ∈ Dom(v) where v(α) = u ;
          if Marked( v )
            then begin
              unifs1 := ∅;
              noRootRW := true
            end
            else begin
              f := Root(v);
              n := Arity(f);
              θ := [f(y₁,...,yₙ)/u];        { where the yᵢ are new variables }
              unifs1 := θ ∘ E-Unifs(θ(u), θ(v), true, true);
            end
        end {else}
    end {then}
  else unifs1 := ∅;
```

{ *Case 2:* *Find unifiers which rewrite u and v at the root and collect in unifs2* }

```
if Id ∈ unifs1 or noRootRW
    then unifs2 := ∅
    else begin
      if Variable(u)
        then Swap(u, v);
      unifs2 := ∅;
      for each (l ≐ r) ∈ E ∪ E⁻¹ do
        for each θ ∈ E-Unifs(u, l, false, true ) do
          unifs2 := unifs2 ∪ θ ∘ E-Unifs(θ(r), θ(v), false, false );
    end;

  currDepth := currDepth - 1;
  return( unifs1 ∪ unifs2 )
end; { E-Unifs }
```

Appendix 2

Huet's Unification Procedure

The basic idea of the higher order unification procedure, as developed by Huet in [72], is to search for unifiers of two lambda-terms one substitution at a time by alternately decomposing terms and finding matching substitutions for the heads, stopping when the subterms are found to be either trivially unifiable, or not unifiable. More specifically, the procedure generates a tree (of OR branches) from a root consisting of the original pair of terms, whose nodes are disagreement sets of pairs of terms not yet unified, and whose arcs are labelled by substitutions found and applied to generate new descendants. The tree is explored and unifiers incrementally created by decomposing pairs of terms until their heads are no longer equal and then finding substitutions which match the heads of pairs, if possible. Identical pairs of terms are fully decomposed and eventually removed from the disagreement set. When either a trivially unifiable disagreement set, composed only of flexible-flexible pairs, is found (success) or an un-unifiable pair, i.e., a rigid-rigid pair with dissimilar top function symbols, is found (failure), a branch is terminated. In general this process may not terminate, since whether two lambda terms are unifiable is only semi-decidable.

We now present a pseudo-Pascal version of Huet's procedure for pre-unifying two terms in the $\alpha\beta\eta$–calculus. The interested reader should consult [71] or [72] for further details.

global variable   T : searchTree;

procedure LambdaUnifiers(e_1, e_2 : λ-terms);
{ This procedure enumerates a complete set of pre-unifiers for
   two λ-terms of the same type. }
var
   N, N' : treeNodes; e_1', e_2' : λ-terms; Σ : substSet; σ, ρ, θ : unifier;
begin
   T := the one node tree consisting of Simplify($\{\{(e_1, e_2)\}\}$);
   while exists an unmarked leaf node N in **T** do
      begin
         Pick some *flexible-rigid* pair $(e_1, e_2) \in N$;
         Σ := Match(e_1, e_2, $FV(N)$);
         if $\Sigma = \emptyset$
            then mark N with "F"
            else
               for each $\sigma \in \Sigma$ **do**
                  begin
                     N' := Simplify($\sigma(N)$);
                     Add a descendant arc from N to N' labelled by σ;
                     if N' is labelled "S"
                        then begin
                           $\theta := Id$;
                           for each ρ on path from N' to root of T **do**
                              $\theta := \rho \circ \theta$;
                           Output(θ)
                        end
                  end
      end
end.

function Simplify(N : disSet) : node;
{ Takes a disagreement set of pairs of terms of the same type and returns
   a node marked with "F" or "S", or a new disagreement set containing
   at least one *flexible-rigid* pair. }
begin
   { Dissolve all *rigid-rigid* pairs. }
   while exists *rigid-rigid* pair (e_1, e_2) in N **do**
      begin
         { Suppose $e_1 = \lambda x_1 \ldots x_n. @_1(e_1^1, \ldots, e_{p_1}^1)$
                 and $e_2 = \lambda y_1 \ldots y_n. @_2(e_1^2, \ldots, e_{p_2}^2)$ }
         { See if heads same. }
         if not $(\lambda x_1 \ldots x_n. @_1 \leftrightarrow_\alpha \lambda y_1, \ldots, y_n. @_2)$
            then Return(N marked with "F");
         { Else we know $\tau(@_1) = \tau(@_2)$ and thus $p_1 = p_2$ }
         Replace (e_1, e_2) by the pairs
            $(\lambda x_1 \ldots x_n. e_i^1, \lambda y_1 \ldots y_n. e_i^2)$ for $1 \le i \le p_1$
      end;
   { Orient pairs. }
   while exists *rigid-flexible* pair $(e_1, e_2) \in N$ **do**
      Replace (e_1, e_2) by (e_2, e_1);
   if exists some *flexible-rigid* pair in N
      then Return(N)
      else Return(N marked with "S")
end;

function Match(e_1, e_2 : λ−terms; V : setOfVars) : substSet;
{ Returns a set of substitutions which matches head of e_1 to head of e_2.
   e_1 is a flexible term $\lambda x_1 \ldots x_n . F(e_1^1, \ldots, e_{p_1}^1)$
   and e_2 is a rigid term $\lambda y_1 \ldots y_n . @(e_1^2, \ldots, e_{p_2}^2)$,
   where $\tau(e_1) = \tau(e_2) = \alpha_1, \ldots, \alpha_n \to \beta$. The set of unifiers
   returned is obtained by *imitating* the head of e_2 and
   by *projecting* e_1 on each of its arguments which preserves the type. }
var Σ : substSet; i : integer;
begin
   { Imitate heading of e_2 if possible. }
   if Constant($@$)
      then $\Sigma := \{\ [\lambda z_1 \ldots z_{p_1} . @(G_1(z_1, \ldots, z_{p_1}), \ldots, G_{p_2}(z_1, \ldots, z_{p_1}))/F]\ \}$;
         { Where $\tau(z_i) = \tau(c_i^1)$, for $1 \leq i \leq p_1$, and the G_j are
            variables not in V such that $\tau(G_j) = \tau(e_1^1), \ldots, \tau(e_{p_1}^1) \to \tau(e_j^2)$}
      else $\Sigma := \emptyset$;
   { Next project F on each of its arguments which has appropriate type. }
   for $i := 1$ **to** p_1 **do**
      if $\tau(e_i^1) = \gamma_1, \ldots, \gamma_{m_i} \to \beta$ for some γ_j { Note that possibly $m_i = 0$. }
         then
            $\Sigma := \Sigma \cup \{\ [\lambda z_1 \ldots z_{p_1} . z_i(H_1^i(z_1, \ldots, z_{p_1}), \ldots, H_{m_i}^i(z_1, \ldots, z_{p_1}))/F]\ \}$;
               { Where $\tau(z_i) = \tau(e_i^1)$ for $1 \leq i \leq p_1$
                  and the H_j^i for $1 \leq j \leq m_i$ are variables not in V
                  of type $\tau(H_j^i) = \tau(e_1^1), \ldots, \tau(e_{p_1}^1) \to \tau(\gamma_j)$}
   Return(Σ)
end;

Appendix 3

HERBRAND'S UNIFICATION ALGORITHM

It is remarkable that in his thesis, Herbrand gave all the steps of a unification algorithm based on transformations on systems of equations. This occurs in Chapter 5 of the thesis, entitled "Properties of True Propositions." Property A concerns whether the matrix of a formula is a "normal identity," which is roughly determined by the presence of *mated pairs* (in the terminology of [4]). The method Herbrand gives for finding such pairs involves finding a set of "associated equations" for a pair of atomic formulae, which are basically the same as our solved pairs. The following passage, from [64], p.148, gives Herbrand's algorithm:

Now, to find an appropriate set of associated equations is easy, if such a set exists; it suffices, for each system of equations between arguments, to proceed by recursion, using one of the following procedures which simplify the system of equations to be satisfied.

(1) If one of the equations to be satisfied equates a restricted variable x to an individual, either this individual contains x, and then the equations cannot be satisfied, or else the individual does not contain x, and then the equation will be one of the associated equations that we are looking for; in the other equations to be satisfied we replace x by the individual;

(2) If one of the equations to be satisfied equates a general variable to an individual that is not a restricted variable, the equation cannot be satisfied;

(3) If one of the equations to be satisfied equates $f_1(\varphi_1, \varphi_2, \ldots, \varphi_n)$ to $f_2(\psi_1, \psi_2, \ldots, \psi_n)$, either the elementary functions f_1 and f_2 are different, and then the equation cannot be satisfied, or they are the same, and then we turn to those equations that equate the φ_i to the ψ_i.

Therefore, if we successively consider each prenex form of P, we shall be able, after a finite and determinate number of steps, to decide whether the proposition P is a normal identity.

BIBLIOGRAPHY

[1]   Ait-Kaci, H., "An Algorithm for Finding a Minimal Recursive Path Ordering," Theoretical Informatics 19:4 (1985) 359-382.

[2]   Anderson, R., "Completeness Results for E-resolution," Proceedings of the AFIPS Spring Joint Computer Conference 36 (1970) 653-656.

[3]   Andrews, P., "Resolution in Type Theory," JSL 36:3 (1971) 414-432.

[4]   Andrews, P., "Theorem Proving via General Matings," JACM 28:2 (1981) 193-214.

[5]   Andrews, P., D. Miller, E. Cohen, F. Pfenning, "Automating Higher-Order Logic," Contemporary Mathematics 29 (1984) 169-192.

[6]   Andrews, P., An Introduction to Mathematical Logic and Type Theory: To Truth Through Proof, Academic Press, Inc. (1986).

[7]   Baader, F., "Characterizations of Unification Type Zero," Proceedings of Third RTA, Chapel Hill, NC (1989), Springer LNCS vol. 355, N. Dershowitz (Ed.), pp. 2–14

[8]   Bachmair, L., Proof Methods for Equational Theories, Ph.D thesis, University of Illinois, Urbana Champaign, Illinois (1987).

[9]   Bachmair, L., Dershowitz, N., and Hsiang, J., "Orderings for Equational Proofs," In Proc. Symp. Logic in Computer Science, Boston, Mass. (1986) 346-357.

[10] Bachmair, L., and Dershowitz, N., "Critical Pair Criteria for the Knuth-Bendix Completion Procedure," unpublished.

[11] Bachmair, L., and Dershowitz, N., "Commutation, Transformation, and Termination," unpublished.

[12] Bachmair, L., Dershowitz, N., and Plaisted, D., "Completion without Failure," Proceedings of CREAS, Lakeway, Texas (May 1987), also submitted for publication.

[13] Bachmair, L., Plaisted, D.A., "Termination Orderings for Associative Commutative Rewriting Systems," JSC 1 (1985) 329-349.

[14] Bachmair, L. and H. Ganzinger, "On Restrictions of Ordered Paramodulation with Simplification," Proceedings of CADE 1990, Kaiserslautern, Germany.

[15] Bachmair, L. and H. Ganzinger, "Rewrite-Based equational Theorem Proving With Selection and Simplification," submitted.

[16] Barendregt, H.P., The Lambda Calculus, North-Holland (1984).

[17] Baxter, L.D., "A Practically Linear Unification Algorithm," Research Report CS-76-13, Dept. of Applied Analysis and Computer Science, Univ. of Waterloo, Waterloo, Ontario, Canada.

[18] Benanav, D., Kapur, D., and Narendran, P., "Complexity of Matching Problems," JSC 3 (1987) 203-216.

[19] Bibel, W., "On Matrices with Connections," JACM **28** (1981) 633–645.

[20] Birkhoff, G., "On the Structure of Abstract Algebras," Proceedings of the Cambridge Philosophical Society, 31:433 (1935).

[21] Brand, D., "Proving Theorems with the Modification Method," SIAM Journal of Computing 4:4 (1975) 412-430.

[22] Bosco, P.G., E. Giovanetti, C. Moiso, "Refined Strategies for Semantic Unification," Proceedings of the International Joint Conference on Theory and Practice of Software Development, Springer LNCS vol. 250 (1987) 276–290.

[23] Breazu-Tannen, V., and Gallier, J., "Polymorphic Rewriting Conserves Algebraic Strong Normalization and Confluence," ICALP 1989 (journal version to appear). LICS 1988, Edinburgh, Scotland.

[24] Buchberger, B., "History and Basic Features of the Critical-Pair/Completion Procedure," JSC 3:1 (1987) 3-38.

[25] Bürckert, H., "Lazy E-Unification - A Method to Delay Alternative Solutions," Workshop on Unification, Val D'Ajol, France (1987).

[26] Bürckert, H., "Some relationships between unification, restricted unification and matching," In J. Siekmann, editor, *Proceedings 8th Conference on Automated Deduction, Oxford*, Springer Verlag, Oxford (England) (1986) 514-524.

[27] Champeaux, D. de, "About the Paterson–Wegman Linear Unification Algorithm," JCSS **32** (1986) 79-90.

[28] Church, A., "A Formulation of the Simple Theory of Types," JSL 5 (1940) 56-68.

[29] Colmerauer, A., Kanoui, H., Pasero, R., and Roussel, P., "Un Systeme de Communication Homme-Machine en Francais," Groupe D'Intelligence Artificielle, U.E.R. de Luminy, Universite D'Aix-Marseille, Luminy, 1972.

[30] Corbin, J., and Bidoit, M., "A Rehabilitation of Robinson's Unification Algorithm," Information Processing 83 (1983) 909-914.

[31] Darlington, J.L., "A Partial Mechanization of Second-Order Logic," Machine Intelligence 6 (1971) 91-100.

[32] Davis, M., and H. Putnam, "A Computing Procedure for Quantification Theory," JACM (March, 1960) 201–215.

[33] Dershowitz, N., "A Note on Simplification Orderings," Information Processing Letters 9:5 (1979) 212-215.

[34] Dershowitz, N., and Manna Z., "Proving Termination with Multiset Orderings," CACM 22 (1979) 465-476.

[35] Dershowitz, N., "Orderings for Term-Rewriting Systems," TCS 17:3 (1982) 279-301.

[36] Dershowitz, N., "Termination of Rewriting," Journal of Symbolic Computation 3 (1987) 69-116.

[37] Dershowitz, N., "Completion and its Applications," CREAS Symposium (1987).

[38] Digricoli, V., and Harrison, M., "Equality-Based Binary Resolution," JACM 33:2 (1986) 253-289.

[39] Dougherty, D., and P. Johann, "An Improved General E-Unification Method," Proceedings of the Tenth CADE, Springer LNCS vol. 449 (1990) 261–275. (Journal version to appear in JSC.)

[40] Dougherty, D., "Higher-Order Unification via Combinators," presented at the Fourth Annual Workshop on Unification, Leeds, UK (1990), submitted.

[41] Eder, E., "Properties of Substitutions and Unifications," JSC 1 (1985) 31-46.

[42] Elliot, C., and Pfenning, F., "A Family of Program Derivations for Higher-Order Unification," Ergo Report 87-045, CMU, November 1987.

[43] Fages, F. and Huet, G., "Complete Sets of Unifiers and Matchers in Equational Theories," TCS 43 (1986) 189-200.

[44] Farmer, W., *Length of Proofs and Unification Theory*, Ph.D. Thesis, University of Wisconsin—Madison (1984).

[45] Farmer, W. "A Unification Algorithm for Second-Order Monadic Terms," Unpublished Technical Report, MITRE Corporation, Bedford, MA.

[46] Fay, M., "First-order Unification in an Equational Theory," Proceedings of the 4th Workshop on Automated Deduction, Austin, Texas (1979).

[47] Felty, A., and Miller, D., "Specifying Theorem Provers in a Higher-Order Logic Programming Language," Ninth International Conference on Automated Deduction, Argonne, Illinois (1988).

[48] Fribourg, L., "SLOG: A Logic Programming Language Interpreter Based on Clausal Superposition and Rewriting," Proceedings of the 1985 Symposium on Logic Programming, Boston, pp. 172–184.

[49] Gallier, J.H. *Logic for Computer Science: Foundations of Automatic Theorem Proving*, Harper and Row, New York (1986).

[50] Gallier, J.H., and Raatz, S., "Extending SLD-Resolution to Equational Horn Clauses using E-Unification," Technical Report MS-CIS-86-44, University of Pennsylvania (1986).

[51] Gallier, J., P. Narendran, S. Raatz, and W. Snyder, "Theorem Proving Using Equational Matings and Rigid E-Unification," to appear in Journal of Association for Computing Machinery (1991).

[52] Gallier, J., P. Narendran, D. Plaisted, W. Snyder, "Rigid E-Unification: NP-completeness and Applications to Theorem Proving," Information and Computation **87** (1990) 129—195 (special issue of selected papers from LICS 1988).

[53] Gallier, J.H., and Snyder, W., "A General Complete E-Unification Procedure," RTA, Bordeaux, 1987.

[54] Gallier, J.H., and Snyder, W., "Complete Sets of Transformations for General E-Unification," to be published in TCS, 1989.

[55] Gallier, J., and W. Snyder, "Designing Unification Procedures using Transformations: A Survey," Bulletin of the EATCS, Number **40** February (1990).

[56] Gallier, J., and W. Snyder, "Inference Systems for Unification: A Survey," to appear in Proceedings of the MSRI Workshop in *Logic From Computer Science*, Springer Verlag (1991).

[57] Goguen, J.A., Meseguer, J., "Eqlog: Equality, Types, and Generic Modules for Logic Programming," Journal of Logic Programming 2 (1984) 179-210.

[58] Goldfarb, W., "The Undecidability of the Second-Order Unification Problem," TCS 13 (1981) 225-230.

[59] Gould, W.E., *A Matching Procedure for Omega-Order Logic*, Ph.D. Thesis, Princeton University, 1966.

[60] Guard, J.R., "Automated Logic for Semi-Automated Mathematics," Scientific Report 1, AFCRL 64-411, Contract AF 19 (628)-3250 AD 602 710.

[61] Guard, J., Oglesby, J., and Settle, L., "Semi-Automated Mathematics," JACM 18 (1969) 42-62.

[62] Hannan, J. and Miller, D., "Enriching a Meta-Language with Higher-Order Features," Workshop on Meta-Programming in Logic Programming, Bristol (1988).

[63] Hannan, J. and Miller, D., "Uses of Higher-Order Unification for Implementing Program Transformers," Fifth International Conference on Logic Programming, MIT Press (1988).

[64] Herbrand, J., "Sur la Théorie de la Démonstration," in: *Logical Writings*, W. Goldfarb, ed., Cambridge, 1971.

[65] Hindley, J.R., Lercher, B., and Seldin, J.P., *Introduction to Combinatory Logic*, London Mathematical Society Lecture Note Series 7, Cambridge University Press (1972).

[66] Holldobler, S., "A Unification Algorithm for Confluent Theories," ICALP (1987).

[67] Holldobler, S., *Equational Logic Programming, Springer-Verlab Lecture Notes in Artificial Intelligence*, Berlin (1989).

[68] Hsiang, J., and M. Rusinowitch, "Proving Refutational Completeness of Theorem Proving Strategies: The Transfinite Semantic Tree Method," to appear in JACM (1991).

[69] Huet, G., "A Mechanization of Type Theory," Proceedings of the Third International Joint Conference on Artificial Intelligence (1973) 139-146.

[70] Huet, G., "The Undecidability of Unification in Third-Order Logic," Information and Control 22 (1973) 257-267.

[71] Huet, G., "A Unification Algorithm for Typed λ-Calculus," TCS 1 (1975) 27 - 57.

[72] Huet, G., *Résolution d'Equations dans les Langages d'Ordre* $1, 2, \ldots, \omega$, Thèse d'Etat, Université de Paris VII (1976).

[73] Huet, G., "Confluent Reductions: Abstract Properties and Applications to Term Rewriting Systems," JACM 27:4 (1980) 797-821.

[74] Huet, G., and Lang, B., "Proving and Applying Program Transformations Expressed with Second-Order Patterns," Acta Informatica 11 (1978) 31-55.

[75] Huet, G., and Lankford, D.S., "On the Uniform Halting Problem for Term Rewriting Systems," Rapport Laboria 283, IRIA, Mars (1978).

[76] Huet, G., "A Complete Proof of Correctness of the Knuth and Bendix Completion Algorithm," JCSS 23 (1981) 11-21.

[77] Huet, G. and Oppen, D. C., "Equations and Rewrite Rules: A Survey," in *Formal Languages: Perspectives and Open Problems*, R. V. Book (ed.), Academic Press, NY (1982).

[78] Hullot, J.-M., "Canonical Forms and Unification," Proceedings CADE-5 (1980) 318-334.

[79] Hussmann, H., "Unification in Conditional Equational Theories," Proceedings of the EUROCAL 1985, Springer Lecture Notes in Computer Science 204, p. 543-553.

[80] Josephson, A., and N. Dershowitz, "An Implementation of Narrowing: The RITE Way," Proceedings of the 1986 Symposium on Logic Programming, Salt Lake City, Utah, pp. 187-197.

[81] Johann, P., *Complete Sets of Transformations for Unification Problems*, PhD. Dissertation, Department of Mathematics, Wesleyan University, Middletown, Conn. (1991).

[82] Jouannaud, J.P., Kirchner, C., and Kirchner, H., "Incremental Construction of Unification Algorithms in Equational Theories," in *Proceedings of the International Conference on Automata, Languages, and Programming*, Lecture Notes in Computer Science, Springer Verlag, Barcelona, 154 (1983) 361-373.

[83] Jouannaud, J.-P., and C. Kirchner, "Solving Equations in Abstract Algebras, a Rule-Based Survey of Unification," in *Festschrift for Robinson*, J.-L. Lassez and G. Plotkin, (Eds.), MIT Press, Cambridge, MA (1991).

[84] Kamin. S., and Levy, J.-J., "Two Generalizations of the Recursive Path Ordering," unpublished note, Department of Computer Science, University of Illinois, Urbana, IL.

[85] Kaplan, S., "Fair Conditional Term Rewriting Systems: Unification, Termination, and Confluence," Technical Report 194, Universite de Paris-Sud, Centre D'Orsay, Laboratoire de Recherche en Informatique (1984).

[86] Kapur, D., Narendran, P. "NP-completeness of the Set Unification and Matching Problems," Proceedings of 8th CADE (1986).

[87] Kapur, D., Narendran, P. "Associative-Commutative Unification is NP-complete," Technical Report, GE Corporate Research and Development Center, Schenectady, NY.

[88] Kapur, D., Narendran, P. "Matching, Unification, and Complexity," Technical Report, GE Corporate Research and Development Center, Schenectady, NY.

[89] Kfoury, A., J. Tiuryn, and P. Urzyczyn, "The Undecidability of the Semi-Unification Problem," Proceedings of STOC (1990).

[90] Kirchner, C., "A New Equational Unification Method: A Generalization of Martelli-Montanari's Algorithm," *Proceedings of the Seventh International Conference on Automated Deduction*, Napa Valley (1984).

[91] Kirchner, C., *Méthodes et Outils de Conception Systematique d'Algorithmes d'Unification dans les Theories Equationnelles*, Thèse d'Etat Université de Nancy I (1985).

[92] Kirchner, C., "Computing Unification Algorithms," LICS'86, Cambridge, MA (1986).

[93] Kirchner, C. and Kirchner, H., *Contribution à la Resolution d'Equations dans les Algèbres Libres et les Varietés Equationnelles d'Algèbres*, Thèse de 3ᵉ cycle, Université de Nancy I (1982).

[94] Knuth, D.E. and Bendix, P.B., "Simple Word Problems in Univeral Algebras," in: Leech, J. (ed.), *Computational Problems in Abstract Algebra*, Pergamon Press (1970) 263-297.

[95] Kowalski, R., "Predicate Logic as Programming Language," Information Processing 74 (1974) 569-574.

[96] Kowalski, R.A., "Algorithm = Logic + Control," CACM **22** : 7 (1979) 424-435.

[97] Kozen, D., "Complexity of Finitely Presented Algebras," Technical Report TR 76-294, Department of Computer Science, Cornell University, Ithaca, New York (1976).

[98] Kozen, D., "Positive First-Order Logic is NP-Complete," IBM Journal of Research and Development, 25:4 (1981) 327-332.

[99] Krishnamoorthy, M.S., and Narendran, P., "On Recursive Path Ordering," TCS 40 (1985) 323-328.

[100] Lankford, D.S., "Canonical Inference," Report ATP-32, University of Texas (1975),

[101] Lankford, D., and Ballantyne, A.M., "The Refutation Completeness of Blocked Permutative Narrowing and Resolution," Proceedings of the Fourth CADE, Austin, Texas (1979).

[102] Lankford, D.S., "On Proving Term Rewriting Systems are Noetherian," Report MTP-3, Mathematics Department, Louisiana Tech University (1979).

[103] Lassez, J., Maher, M., and Marriott, K., "Unification Revisited," Report RC 12394 (55630), IBM, T.J. Watson Research Center, Yorktown Heights (1986).

[104] Levy, A., *Basic Set Theory*, Springer-Verlag, New York (1979).

[105] Lindstrom, G., "Functional Programming and the Logical Variable," ACM Symposium on Principles of Programming Languages (1985).

[106] LLoyd, J.W., *Foundations of Logic Programming*, Springer-Verlag, New York, 1984.

[107] Lucchesi, C.L., "The Undecidability of the Unification Problem for Third Order Languages," Report CSRR 2059, Dept. of Applied Analysis and Computer Science, University of Waterloo (1972).

[108] Machtey, M. and Young, P.R., *An Introduction to the General Theory of Algorithms*, Elsevier North-Holland, NY (1977).

[109] Martelli, A., Montanari, U., "An Efficient Unification Algorithm," ACM Transactions on Programming Languages and Systems 4:2 (1982) 258-282.

[110] Martelli, A., Montanari, U., "Unification in Linear Time and Space: A Structured Presentation," Internal Report B76-16, Ist. Di Elaborazione della Informazione, Consiglio Nazionale delle Ricerche, Piza, Italy, July 1976.

[111] Martelli, A., Moiso, C., Rossi, G.F., "An Algorithm for Unification in Equational Theories," Third Conference on Logic Programming, Utah (1986).

[112] Miller, D., *Proofs in Higher-Order Logic*, PhD. Dissertation, CMU (1983).

[113] Miller, D., and Nadathur, G., "Higher-Order Logic Programming," Proceedings of the Third International Conference on Logic Programming, London (1986).

[114] Miller, D., and Nadathur, G., "A Logic Programming Approach to Manipulating Formulas and Programs," IEEE Symposium on Logic Programming, San Franciso (1987).

[115] Miller, D., and Nadathur, G., "Some Uses of Higher-Order Logic in Computational Linguistics," 24th Annual Meeting of the Association for Computational Linguistics (1986) 247—255.

[116] Milner, R., "A Theory of Type Polymorphism in Programming," Journal of Computer and Systems Sciences 17:3 (1978) 348-375.

[117] Morris, J.B., "E-Resolution: Extension of Resolution to Include the Equality Relation," IJCAI (1969).

[118] Nadathur, G., "A Higher-Order Logic as the Basis for Logic Programming," Ph.D. Dissertation, Department of Computer and Information Science, University of Pennsylvania (1986).

[119] Nipkow, T., and Z. Qian, "Modular Higher-Order E-Unification," Proceedings of Fourth RTA, Como, Italy (1991), Springer LNCS vol. 488, R. Book (Ed.), pp. 200–214.

[120] Nutt, W., Réty, P., Smolka, G., "Basic Narrowing Revisited," SEKI Report SR-87-07, Universität Kaiserslautern, West Germany, July 1987.

[121] Padawitz, P., *Computing in Horn Theories*, EATCS Monographs on Theoretical Computer Science, Vol. 16, Springer-Verlag, Berlin (1989).

[122] Paterson, M.S., Wegman, M.N., "Linear Unification," Journal of Computer and System Sciences 16 (1978) 158-167.

[123] Paulson, L.C., "Natural Deduction as Higher-Order Resolution," Journal of Logic Programming 3:3 (1986) 237-258.

[124] Peterson, G., "A Technique for Establishing Completeness Results in Theorem Proving with Equality," SIAM Journal of Computing 12:1 (1983) 82-100.

[125] Pfenning, F., *Proof Transformations in Higher-Order Logic*, Ph.D. thesis, Department of Mathematics, Carnegie Mellon University, Pittsburgh, Pa. (1987).

[126] Pfenning, F., "Partial Polymorphic Type Inference and Higher-Order Unification," in *Proceedings of the 1988 ACM Conference on Lisp and Functional Programming*, ACM, July 1988.

[127] Pfenning, F., and Elliott, C., "Higher-Order Abstract Syntax," to appear in *Proceedings of the SIGPLAN '88 Symposium on Language Design and Implementation*, ACM, June 1988.

[128] Pietrzykowski, T., "A Complete Mechanization of Second-Order Logic," JACM 20:2 (1971) 333-364.

[129] Pietrzykowski, T., and Jensen, D., "Mechanizing ω-Order Type Theory Through Unification," TCS 3 (1976) 123-171.

[130] Plaisted, D., and S. Greenbaum, "Problem Representations for Back Chaining and Equality in Resolution Theorem Proving," Proceedings of the First Conference on Artificial Intelligence Applications (1984).

[131] Plotkin, G., "Building in Equational Theories," Machine Intelligence 7 (1972) 73-90.

[132] Prawitz, D., "An Improved Proof Procedure," *Theoria* **26** (1960) 102-139.

[133] Raatz, S., *Aspects of a Graph-based Proof Procedure for Horn Clauses*, Ph.D. Thesis, University of Pennsylvania (1987).

[134] Reddy, U.S., "Narrowing as the Operational Semantics of Functional Languages," 1985 IEEE Symposium on Logic Programming, Boston, 138-151.

[135] Rety, P., Kirchner, C., Kirchner, H., and Lescanne, P., "Narrower: A New Algorithm for Unification and its Application to Logic Programming," Proceedings of the First RTA Conference, Dijon, France (1985).

[136] Rety, P., "Improving Basic Narrowing Techniques," Proceedings of the Second RTA Conference, Bordeaux, France (1987).

[137] Robinson, G.A., Wos, L., Carson, D.F., and Shalla, L. "The Concept of Demodulation in Theorem Proving," JACM 14 (1967) 698-709.

[138] Robinson, G. A. and Wos, L., "Paramodulation and Theorem-Proving in First-order Logic with Equality," Machine Intelligence 4 (1969) 135-150.

[139] Robinson, J.A., "A Machine Oriented Logic Based on the Resolution Principle," JACM 12:1 (1965) 23 -41.

[140] Robinson, J.A., "A Review on Automatic Theorem Proving," Annual Symposia in Applied Mathematics, 1-18 (1967).

[141] Robinson, J.A., "Mechanizing Higher-Order Logic," Machine Intelligence 4 (1969) 151-170.

[142] Robinson, J.A., "Computational Logic: The Unification Computation," Machine Intelligence 6 (1971) 63-72.

[143] Rusinowich, M., *Démonstration Automatique: Techniques de Réécriture*, InterEditions, Paris (1989).

[144] Sakai, K., *On Computer Aided Mathematical Reasoning*, Technical Report, Institute for New Generation Computer Technology.

[145] Siekmann, J. H., "Universal Unification," Proceedings of the Seventh CADE, Napa (1984) 1-42.

[146] Slagle, J.R., "Automated Theorem Proving for Theories with Simplifiers, Commutativity, and Associativity," JACM 21 (1974) 622-642.

[147] Shortliffe, E.H., *Computer-Based Medical Consultation: MYCIN*, Elsevier North-Holland, New York, 1976.

[148] Snyder, W., and J. Gallier, "Higher-Order Unification Revisited: Complete Sets of Transformations," Journal of Symbolic Computation 8 (1989) 101— 140; reprinted in [154], pp. 565-604.

[149] Snyder, W. and C. Lynch, "Complete Inference Systems for Horn Clause Logic with Equality: A Foundation for Logic Programming with Equality," Proceedings of Second International Workshop on Conditional and Typed Rewriting Systems, Montreal, Canada (1990). (Journal version submitted.)

[150] Snyder, W., "Higher-Order E-Unification," Proceedings of Tenth CADE, Kaiserslautern, Germany (1990), Springer LNCS vol. 449, M. Stickel (Ed.), pp. 573-587.

[151] Snyder, W. and C. Lynch, "Goal Directed Strategies for Paramodulation," Proceedings of RTA, Como, Italy (1991), *Springer Verlag Lecture Notes in Computer Science*, Vol. 488, R. Book (ed.), pp. 150-161. (Journal version submitted.)

[152] Snyder, W., and C. Lynch, "Basic Paramodulation," in preparation.

[153] Statman, R., "On the Existence of Closed Terms in the Typed λ-Calculus II: Transformations of Unification Problems," TCS 15:3 (1981) 329-338.

[154] *Unification*, C. Kirchner (ed.), Academic Press, San Diego (1990).

[155] Venturini Zilli, M., "Complexity of the Unification Algorithm for First-Order Expressions," Calcolo 12:4 (1975) 361-372.

[156] Winterstein, G., "Unification in Second-Order Logic," Electronische Informationsverarbeitung und Kybernetik 13 (1977) 399-411.

[157] Wos, L., G. Robinson, and D. Carson, "Efficiency and Completeness of the Set of Support Strategy in Theorem Proving," JACM **12**:4 (Oct. 1965) 536-541.

[158] You, J.-H., Subrahmanyam, P.A., "A Class of Confluent Term Rewriting Systems and Unfication," Journal of Automated Reasoning **2** : 4 (1986) 391-418.

[159] Zaionc, M., "The Set of Unifiers in Typed λ-Calculus as a Regular Expression," Proceedings of the RTA 1985.

[160] Zhang, H., *Reduction, Superposition, and Induction: Automated Reasoning in an Equational Logic*, Ph.D. Thesis, Rensselaer Polytechnic Institute (1988).

Progress in Computer Science and Applied Logic

Progress in Computer Science and Applied Logic is a series that focuses on scientific work of interest to both logicians and computer scientists. Thus both applications of mathematical logic will be topics of interest. An additional area of interest is the foundations of computer science.

The series (previously known as *Progress in Computer Science*) publishes research monographs, graduate texts, polished lectures from seminars and lecture series, and proceedings of focused conferences in the above fields of interest. We encourage preparation of manuscripts in some form of TeX for delivery in camera-ready copy, which leads to rapid publication, or in electronic form for interfacing with laser printers or typesetters.

Proposals should be sent directly to the editors or to:
Birkhäuser Boston, 675 Massachusetts Ave., Cambridge, MA 02139

Progress in Computer Science and Applied Logic